Green Heroes

László Erdős

Green Heroes

From Buddha to Leonardo DiCaprio

 Springer

László Erdős
MTA-DE Lendület Functional and Restoration Ecology Research Group, Debrecen
Institute of Ecology and Botany MTA Centre for Ecological Research
Vácrátót, Hungary

ISBN 978-3-030-31808-6   ISBN 978-3-030-31806-2   (eBook)
https://doi.org/10.1007/978-3-030-31806-2

Cover illustration: Photo by László Balogh

This Springer imprint is published by the registered company Springer Nature Switzerland AG.
The registered company address is: Gewerbestrasse 11, 6330 Cham, Switzerland

To Dolly

Acknowledgements

Writing this book was a challenging yet highly rewarding journey. It was a journey on which I received continuous emotional, technical, and financial support from many sources.

I owe the greatest debt to my late mother, who, despite tremendous personal and financial difficulties, created the happiest and most democratic family environment I can possibly imagine. She fostered my interest in animals and the natural world and helped me find ways to become an activist. My mother taught history and literature at a secondary school but also was a great animal lover and an amateur naturalist. Her love and wisdom are missed every minute of every day.

I am especially grateful to my wife, Dolly, for her invaluable comments and suggestions on the manuscript. I also thank her for her love and patience. Without her full support and encouragement, I would never have been able to write this book.

I would like to thank my father and brother for their assistance during this and my earlier projects.

I am grateful to Mike Cooke from Ambios Ltd for his linguistic advice while finalising the manuscript. Special thanks are due to Simon Roper, the Director of Ambios Ltd, for his support in preparing the book.

I owe a great deal to Dr. László Körmöczi, Dr. György Kröel-Dulay, Dr. Péter Török, Dr. Zoltán Botta-Dukát, Dr. László Garamszegi, and Dr. András Takács-Sánta, who made it possible for me to focus on my book and allowed to develop and teach university courses on the subject.

For their useful comments on how to improve the manuscript, I want to express my gratitude to Dr. Zoltán Bátori, Dr. Judit Bódis, Viktória Cseh, Dr. János Csiky, Mira Devánszki, Dr. Erika Farkas Csamangó, Viktória Füzesi, Dr. László Gallé, Dr. Róbert Gallé, Krisztina Gellény, Dr. Balázs Kevey, Dr. Martin Magnes, Dr. György Málovics, Dr. Katalin Margóczi, Dr. Zsolt Molnár, Dr. Róbert Pál, Dr. Zsolt Palkó, Dr. Károly Penksza, Dr. Stephen Puryear, Dr. Petr Smýkal, Dr. János I. Tóth, and Dr. Csaba Tölgyesi.

My work was supported by the National Youth Excellence Scholarship, which made me possible to obtain materials for writing this book.

Last but not least, I want to express appreciation to all green heroes, those who are mentioned in this book and those who are not, for their tireless and unselfish work to protect animals, the natural world, and our environment.

Contents

Introduction   1

Part I   Heroes for Animals

Do Not Kill – Buddha's Compassion for Animals   7

Saint Francis of Assisi – The Patron Saint of Animals 11

Saint Francis of Paola – Humility and Non-violence 15

Animals Are Back – The Ethics of Jeremy Bentham 19

Pioneers of Animal Advocacy in the Nineteenth and the Early
Twentieth Centuries 23

Charles Darwin – Evolution and the Fall of Human Superiority 27

Reverence for Life – Albert Schweitzer's Biocentrism 33

Cleveland Amory and the Fund for Animals 39

Spokesmen for Animals – Richard Ryder, Peter Singer,
and Tom Regan 43

How Henry Spira Put Animal Liberation into Practice 49

Ingrid Newkirk, Alex Pacheco, and PETA 55

Talking Apes – Ambassadors of the Animal Kingdom
in the Human World 61

Parrots, Dolphins, Seals – And What They Have Taught Us 67

In the Front Line of Animal Advocacy – From Brigitte Bardot
to Lek Chailert 71

Part II   Heroes for Nature

Beginnings of the Conservation Movement   79

Aldo Leopold and the Land Ethic   85

Under the Banner of the Giant Panda   89

Arne Naess and the Deep Ecology Movement   93

Living with Big Cats – The Story of Joy and George Adamson   97

The Modern Noah – Gerald Durrell's Mission to Save
Endangered Animals.   103

Farley Mowat Never Cried Wolf   107

Jane Goodall – A Lifelong Optimist   111

No One Loved Gorillas More – The Life and Legacy
of Dian Fossey   117

Biruté Galdikas and the People of the Forest.   125

Jacques-Yves Cousteau – In Awe of the Oceans.   129

Mike Pandey – Using the Power of Motion Picture   133

David Attenborough – The Grand Old Man of Natural
History Filmmaking   137

Paul Watson – The Daredevil of Conservation.   143

Part III   Heroes for the Environment

The Environmental Movement Is Born – Rachel Carson
and Silent Spring.   151

Environmentalism Gaining Momentum – Denis Hayes
and Earth Day   155

The Greenpeace-Story   159

Chico Mendes – A Martyr for the Rainforest.   165

Missing in Action – The Story of Bruno Manser   171

Oil Is Blood – The Fight of Ken Saro-Wiwa.   177

A Passion for Trees – Wangari Maathai and the Green
Belt Movement.   181

Vandana Shiva – Defending Traditional Agriculture   187

**Tree Huggers and Hunger Strikers – Environmental
Leadership in India**.   193

The Sacrifice of Berta Cáceres.   199

**Don't Waste It! – Rossano Ercolini, Bea Johnson,
and Ian Kiernan**   203

Al Gore – The Climate Crusader.   207

Arnold Schwarzenegger – The Strong Man of Environmentalism   213

Greening Hollywood – The Activism of Leonardo DiCaprio   217

Epilogue   223

Introduction

What kind of story can start with Buddha and conclude with Leonardo DiCaprio? Well, a story of a movement that unites animal advocates, nature conservationists, and environmentalists, a movement that ranges from an ancient religious leader to a contemporary movie star. In this book I want to tell some of the most important, most exciting, and most inspiring parts of that story.

This book is a collection of my personal heroes. My aim was to write about some of the leading characters in the green movement, who have a global importance, and whose thoughts or actions have a huge motivational power. Instead of conventional biographies, I prepared portraits, based on biographical data, scientific facts, celebrity stories, historical circumstances, ethical principles, green strategies and tactics, and thought-provoking quotes.

The selection is admittedly not objective, nor was it intended as a perfect list. There are many other well-known heroes who would have deserved to be mentioned in this book but could not be included due to limited space. And there are even more heroes who are hardly known but nevertheless work hard to make the world a better place. Thus, my book is not final. I encourage you to add your heroes or compile your own list of green icons!

The personalities you will read about in this book form a quite eclectic group: saints and scientists, philosophers and activists, writers and filmmakers, indigenous leaders and clergymen, movie stars and politicians, women and men, humans and non-humans. I try to see this variety as an asset rather than a drawback. Diversity is of great value not only in the natural world, but also in the green movement. Indeed, this diversity may be the very key for our success!

I arranged the characters into three groups, according to the three main directions of the green movement: animal advocacy, nature conservation, and environmentalism.

Animal advocacy is interested in the rights or well-being of individual animals. Its focus is on individual organisms who have interests, welfare or a quality of life. Animal advocates aim to cease or lessen animal mortality and suffering evoked by humans.

© Springer Nature Switzerland AG 2019
L. Erdős, *Green Heroes*, https://doi.org/10.1007/978-3-030-31806-2_1

Nature conservation's primary concern is not individuals but ecological entities such as populations, species, biotic communities, and ecosystems. Its focus is on preserving natural and near-natural habitats and maintaining the diversity of life at several levels (e.g. the genetic diversity of populations, species diversity, and habitat diversity).

I define environmentalism as an activity that aims to optimise the environmental parameters of the human species. Environmentalism is primarily interested in safeguarding the future of the human species, although most activities connected with environmentalism are clearly beneficial to other species as well.

Despite the theoretical separation outlined above, the three branches form a common platform usually referred to as 'green issues.' The membership of the three directions overlap considerably, and all of them reject treating living beings and the natural world as mere resources.

Due to the overlap, some of my heroes were hard to classify into any one of the three categories. The philosophy of Arne Naess could fit equally well into environmentalism as it fits into nature conservation. Dian Fossey can be regarded as a conservationist and an animal advocate at the same time, since she was as much interested in the survival of the mountain gorilla population as she was in the wellbeing of individual gorillas. Many activists such as Jane Goodall or Leonardo DiCaprio are involved in all three fields. To put it briefly, I am sure my present classification is not perfect, but I think it is acceptable for the purposes of this book.

At the end of the chapters I give references for further reading. For each book I tried to provide the bibliographical data of the most recent, easily available edition. I also list key internet addresses and films. At the end of the Introduction there are some general sources which are connected to the topics covered in two or more chapters.

During the writing of this book I learned a lot – not only *about* my heroes, but also *from* them. I hope these great figures will inspire the readers of this book as much as they have inspired myself.

Worth Reading

Bekoff, M., & Meaney, C. A. (Eds.). (2013). *Encyclopedia of animal rights and animal welfare*. Abingdon: Routledge.

Brockington, D. (2009). *Celebrity and the environment: Fame, wealth and power in conservation*. London: Zed Books.

DesJardins, J. R. (2006). *Environmental ethics: An introduction to environmental philosophy*. Belmont: Thomas Wadsworth.

Desmond, K. (2008). *Planet savers: 301 extraordinary environmentalists*. Sheffield: Greenleaf Publishing.

Mazur, L., & Miles, L. (2009). *Conversations with green gurus: The collective wisdom of environmental movers and shakers*. Chichester: Wiley.

McNeil, J. R. (2000). *Something new under the Sun: An environmental history of the twentieth-century world*. New York: W. W. Norton & Company.

Nolt, J. (2015). *Environmental ethics for the long term: An introduction*. New York: Routledge.

Palmer, J. A. (Ed.). (2001). *Fifty key thinkers on the environment*. Abingdon: Routledge.
Strong, D. H. (1988). *Dreamers and defenders: American conservationists*. Lincoln: University of Nebraska Press.
Switzer, J. V. (2003). *Environmental activism: A reference handbook*. Santa Barbara: ABC-Clio.

Worth Watching

A Fierce Green Fire (2012)

Part I
Heroes for Animals

Do Not Kill – Buddha's Compassion for Animals

The history of animal advocacy can be traced back to ancient India, where ahimsa, i.e., non-violence, has been regarded as the highest virtue for at least two and a half millennia. From its very inception, the concept of ahimsa included non-human animals in the moral sphere. Though it would be an exaggeration to claim that the animal advocacy movement was born some 2500 years ago, the views some major thinkers of the era had on animals were surprisingly modern.

Vardamana Mahāvīra, the great character of Jainism, taught that it is morally wrong to harm sentient beings. Non-violence to animals is as important in Jaina faith as non-violence to humans. Indeed, non-violence is the primary ethical rule in Jainism. Accordingly, Jains never harm sentient beings deliberately and take the greatest care not to harm anyone unintentionally. They follow a strict vegetarian diet and do not take part in activities that cause suffering to sentient beings.

Another great religious leader of the time took a similar spiritual path, which also led to non-violence. His name was Gautama Siddhārtha, but today he is commonly known as the Buddha.

Little is known about Buddha's life but his historicity is generally accepted. Gautama Siddhārtha was the eldest son of a ruler in the north of India. The exact dates of his birth and death are unknown; according to many historians he must have lived in the fifth century BCE. Born into a royal family, he had a privileged life yet he became deeply upset by the universality of suffering. Still in his youth he embarked on a spiritual quest. For some years he lived a most austere life but discovered that neither extreme asceticism nor excessive luxury can lead to enlightenment. Tradition has it that, aged 35, he started to meditate under the canopy of a pipal tree and decided not to get up until he had attained enlightenment. After he was enlightened, he became known as Buddha, meaning 'the Enlightened One.' For the next four and a half decades he taught in the northern parts of the Indian subcontinent, where he gathered a considerable number of followers. Buddha is said to have died at the age of 80.

© Springer Nature Switzerland AG 2019
L. Erdős, *Green Heroes*, https://doi.org/10.1007/978-3-030-31806-2_2

Buddha did not write down his teachings. His thoughts were transmitted in oral form and written down centuries after his death. However, it is reasonable to assume that the spirit of Buddhism reflects the teachings of Buddha.

Non-violence is the central moral imperative of Buddhism, while compassion is one of the most basic qualities a Buddhist has to strive for. The fact that non-human animals are sentient beings is taken seriously by Buddhism (which is all the more important when taking into account that animal sentience is usually disregarded or downplayed in western thought). In fact, according to the teachings of Buddha, there is no fundamental difference between humans and animals; for Buddha, the happiness of humans and animals was equally important. From this it follows that compassion is due to humans and non-human animals alike. In practice this means that we should not harm any sentient beings. Buddha emphasised that it is unethical to increase our happiness by inflicting suffering on others, thus, the use of animals for human benefit is not acceptable.

The first and most important precept of Buddhism, 'Do not kill,' applies to all sentient beings: it forbids to kill humans as well as animals. Moreover, Buddhist scriptures repeatedly quote Buddha as explicitly stating that followers of him should not hurt any sentient being. Consequently, jobs that involve the killing or exploitation of sentient beings are prohibited to Buddhists. Hunting, fishing, the capture of wild animals, or taking part in animal agriculture is irreconcilable with Buddhist ethics.

Buddha did not accept any hierarchy of living beings, at least not in the western sense of the word. Humans are no more valuable than animals and we are not entitled to dominate, oppress, or abuse animals based on some arbitrarily selected characteristic such as intelligence or tool-use. All sentient beings deserve compassion and have to be treated accordingly. Buddhism offers a fully egalitarian outlook.

Buddhist ethics requires action. Thus, it is not enough to have compassionate feelings: they have to be put into practice. In Buddhism, real compassion is active: we are obliged to help other sentient beings and relieve their suffering whenever possible.

There is some ambiguity about whether Buddha was a vegetarian. However, it seems logical to assume that he did not eat meat and (with very few and rather special exceptions) expected the same from his followers. The first precept prohibits not only killing, but also indirectly contributing to the killing of sentient individuals. Eating meat involves the killing of a sentient being, therefore, it violates the first precept and is also contrary to compassion. Although not all Buddhists are vegetarians, their proportion is rather high. Plant-based diet is a quite natural part of Buddhist faith and, as a logical expression of non-violence and compassion, abstaining from meat is highly respected even among non-vegetarian Buddhists, particularly in Asia.

In addition to animal advocacy, Buddha's teachings are of great relevance to nature conservation as well. Buddhist scriptures value the beauty of the natural world and usually see forests as ideal environments for meditation. Also, Buddhism does not view humanity as separate from nature but as an integral part of it. Buddhism realises the interconnectedness of everything in nature, which is akin to how ecologists see the biosphere. In a Buddhist understanding, to harm or destroy nature is to harm or destroy ourselves.

Buddhism values simplicity, placing an emphasis on need instead of greed. This is probably more timely today than it was in Buddha's time. Overconsumption is one of the key reasons behind the current destruction of our environment. If consumerist societies were content with a modest lifestyle, the current global ecological crisis could be mitigated considerably.

Buddha was, beyond doubt, one of the most animal-friendly founders of religions in human history. His thoughts have had tremendous influence, and Buddhism has developed into one of the world's major religious groups. Animal advocacy may not be the central topic for all Buddhists, but Buddha's positive attitudes towards animals have had considerable effect on a huge number of his followers. Not only is Buddhism consistent with animal advocacy, but compassion for animals is at the heart of Buddhism. Countless Buddhists, including rulers, religious leaders, and ordinary practitioners, have done everything in their power to help animals. Many individual activists and NGOs continue to work for animals based on their Buddhist faith. Moreover, the teachings of Buddha have the power to enrich the lives of the followers of other religious traditions. Indeed, one needn't be a Buddhist to live a life based on non-violence and compassion.

Worth Reading

Kapleau, P. (1982). *To cherish all life: A Buddhist case for becoming vegetarian.* San Francisco: Harper & Row.

Page, T. (1999). *Buddhism and animals: A Buddhist vision of humanity's rightful relationship with the animal kingdom.* London: UKAVIS Publications.

Phelps, N. (2004). *The great compassion: Buddhism and animal rights.* New York: Lantern Books.

Phelps, N. (2007). *The longest struggle: Animal advocacy from Pythagoras to PETA.* New York: Lantern Books.

Worth Browsing

dharmavoicesforanimals.org
aprayerforcompassion.com

Worth Watching

Animals and the Buddha (2014)
A Prayer for Compassion (2018)

Saint Francis of Assisi – The Patron Saint of Animals

Saint Francis of Assisi is one of the most venerated figures in the history of Christianity. At the same time, he is probably the best known medieval herald of animal advocacy, and his teaching continues to serve as a reference point for today's animal welfare and animal rights activists.

Saint Francis was born in the Italian town of Assisi in either 1181 or 1182. He was baptised as Giovanni (John), but for some reason he was called Francesco (Francis) from his childhood onwards. The son of an affluent cloth merchant and his French wife, Francis lived the life of a spoiled boy. Instead of entering the family business, he longed for military fame. In 1202, in the battle between Assisi and the nearby town of Perugia, Francis was taken prisoner. When he was released after a year, his life began to change completely through a series of watershed experiences.

Francis fell seriously ill, which made him turn inward and re-evaluate his life. One day Francis, overcoming his earlier fear and abhorrence, embraced and kissed a man suffering from leprosy. Another event of key importance happened when Francis was praying in an abandoned chapel outside Assisi. He heard a voice saying 'Francis, go and repair my house, which, as you see, is falling into ruins.'[1] He took the words literally and started to renovate church buildings. He even sold some clothes from the storehouse of his father to finance the renovation works. The ensuing conflict between father and son concluded with Francis taking off all his clothes in the presence of the bishop and a crowd of shocked onlookers. The act symbolised that he did not want to depend on his father's wealth and aimed to live in perfect poverty.

Francis wanted to follow the example of Christ. His sole aim was to live a simple life and to spend time working, praying, preaching, and helping the needy. He was regarded as an insane by many for giving up his privileged life. Others, however, felt great respect for him. Before long Francis attracted an increasing number of admirers. In 1210, accompanied by some of his followers, Francis went to the pope, who

---

[1] Bonaventure: Life of Francis, cited in: Kureethadam, J. I. (2019): The Ten Green Commandments of Laudato Si'. Liturgical Press, Collegeville, p. 1.

© Springer Nature Switzerland AG 2019
L. Erdős, *Green Heroes*, https://doi.org/10.1007/978-3-030-31806-2_3

approved their lifestyle and granted them permission to preach, even though neither Francis nor most of his companions were priests.

Soon Francis was a real star. Wherever he turned up, he was greeted by enthusiastic fans. In 1219 Francis travelled to the Near East. Amidst the raging war between Muslims and Christians, he crossed enemy lines to meet the sultan of Egypt. The two had a friendly discussion, which, given the atmosphere of the time, should be regarded as a notable achievement. Not surprisingly, the sultan did not convert to Christianity, but the fact that Francis returned to the Crusader camp unharmed is itself almost a miracle.

During the absence of Francis, troubles had arisen in the rapidly growing community of his disciples. Francis returned to Italy and prepared the rule of the order. He spent much time praying and contemplating. In the early 1220s, at a church in Greccio, Italy, Francis installed the first nativity scene to remind people of the simple circumstances to which Jesus was born.

In August 1224 Francis ascended Mount La Verna and stayed there in retreat for a couple of weeks. Francis is said to have had an ecstatic vision on the mountain, during which he received the stigmata: the five wounds of Christ appeared on his hands, feet, and side. Throughout the rest of his life Francis did his best to hide the wounds from his followers; only after his death did the stigmata of Francis become widely known.

Due to constant self-mortification, Francis' physical condition was rapidly deteriorating during the 1220s. He died in 1226 and was canonised two years later.

Many have tried to evaluate the role Saint Francis of Assisi played as a forerunner of the green movements in general and animal advocacy in particular. According to historian Lynn White Jr., 'Francis tried to depose man from his monarchy over creation and set up a democracy of all God's creatures.'[2] In Francis' view, the beauty of nature reflects the wisdom of the Creator. Thus, people have to respect and protect the natural world. Moreover, Francis emphasised the kinship of all creatures, as all had been created by the same God. Accordingly, he called all beings his brothers and sisters. Francis loved even those animals that most people find repulsive. He would put earthworms from the road back onto the grass, so that they would not be trampled upon.

The most famous example of the special bond Francis had with animals is his sermon to the birds. According to the legend, Francis approached a flock of birds, greeted them, and was surprised to see that they did not fly away. The birds listened carefully to the words of Francis, who spoke to them about how much God loved them. From that day on, Francis would usually preach to birds and other animals. Another widely known legend tells us about Francis taming a wolf near the town of Gubbio. The wolf had been killing farm animals and even humans, but Francis talked with the wolf and managed to reconcile the two parties, whereafter the inhabitants of Gubbio and the wolf lived in peace.

---

[2] White, Lynn Jr. (1967): The Historical Roots of Our Ecological Crisis. Science 155: p. 1206.

Numerous other legends of Saint Francis testify to his love of animals. On one occasion Brother Francis received a tench from a fisherman. He felt so sorry for the fish that he immediately put the animal back into the water. Whenever he had the opportunity to do so, he released caught fish, warning them to be more careful so that they would not be caught again. No less interesting is the story of Saint Francis and the doves. When he met a boy who had caught some doves, Francis persuaded the boy to give the birds to him and turned to the doves with the following words: 'O my little sisters the doves, so simple, so innocent, and so chaste, why did you allow yourselves to be caught? I will save you from death, and make your nests, that you may increase and multiply, according to the command of God.'[3]

Francis is said to have repeatedly bought lambs to save them from being slaughtered. Moreover, he even envisioned animal welfare regulations and wanted to persuade the emperor to enact a law to protect birds.

It is no wonder that Saint Francis of Assisi is the patron saint of animals and animal advocates. World Animal Day is celebrated on his feast day on October the 4th. On this day, Christians take their pets to church to be blessed and many believers pray for enslaved and abused animals. Also, there are countless secular events about animals and animal advocacy throughout the world on this special holiday.

Saint Francis of Assisi was free of material desires and lived a life as simple as possible. He wore a rough brown tunic tied with a knotted rope, fasted almost all year round, usually slept on the ground, and detested money. It is not necessary that all of us imitate Francis' extreme simplicity, but his way of life certainly has something to teach to most of us who live in consumerist societies, where the continuous desire to earn and spend more and more money is a key factor in the rapid destruction of the biosphere.

Whether someone believes in the literal truth of the legends of Saint Francis, or regards them as nothing more than lovely tales, one thing is certain: Francis must have felt an extraordinary empathy and kinship with animals. His example clearly shows that to love and serve both God and humanity does not preclude loving and serving animals. And this is why Saint Francis of Assisi is rightly considered one of the greatest characters in the history of animal advocacy.

Worth Reading

Linzey, A. (1995). *Animal theology*. Urbana: University of Illinois Press.

---

[3] Quoted in Ugolino: The Little Flowers of St. Francis of Assisi. Christian Classics Ethereal Library, http://www.ccel.org/download.html?url=/ccel/ugolino/flowers.pdf, p. 50.

Saint Francis of Paola – Humility and Non-violence

About two centuries had passed after the death of Saint Francis of Assisi when another great Francis appeared on the stage of history. It was Saint Francis of Paola, the vegan friar, whose destiny was connected to the spirit of Saint Francis of Assisi even before he was conceived.

The parents of Saint Francis of Paola had been married for a while when, afraid of remaining childless, they started to pray to Saint Francis of Assisi, seeking his intercession. When their first child was born, they named him Francis, after the great saint whom they had asked for help. The exact birth date of Francis is debated. According to most sources it was around 1416, but some believe it must have been somewhat later.

Aged 13 the boy was sent to a Franciscan friary for a year. He liked the simple lifestyle of the Franciscans. Not much later, after returning from a pilgrimage with his parents, Francis decided to live in seclusion. He settled in a hut in the proximity of Paola, his hometown. After a while he sought a remoter location, where he spent his time praying and contemplating. His austerity was extreme: he wore a hair shirt, was always barefoot, and refused to touch money.

As news spread about the saintly hermit, more and more people visited him, seeking spiritual advice, or wanting to join him. With the help of his followers and inhabitants from the nearby settlements, Francis built a monastery.

Francis' fame rested not only on his piety and lifestyle, but also on his ability to cure with a mix of words and medicinal herbs. Before long Francis was nicknamed 'the man of herbs.' An increasing number of pilgrims wanted to see him in the hope that Francis would heal them. Francis may have known how to use herbs, but many of his contemporaries were sure that he also had supernatural powers. Francis was widely regarded as a miracle worker.

The reputation of Francis gradually reached several European countries. Even the French king Louis XI, heard about the holy man. When Louis' health was declining, he sent a legation to ask Francis to visit and cure him. But Francis did not want to leave his homeland and his followers. Also, he had always had a strong dislike of the pomp and luxury surrounding important personalities. So, he was

© Springer Nature Switzerland AG 2019
L. Erdős, *Green Heroes*, https://doi.org/10.1007/978-3-030-31806-2_4

reluctant to set off for France. When the king realised that his appeals had been futile, he turned to the pope, who, out of political considerations, ordered Francis to accept the king's invitation. Thus, Francis left Italy, and, after a long journey, arrived in Tour, where Louis resided.

However, the king's wish to become healthy again was not fulfilled. Francis taught him that spiritual healing is more important than bodily healing. Francis comforted him and prepared him for death. Louis died in 1483.

The next king liked Francis' personality so much that he managed to persuade him not to travel home. So, Francis remained in France and never returned to his fatherland. He served partly as an informal papal ambassador, partly as a royal counselor and spiritual guide for Louis' successors. Thus the man whose desire it was to live as a hermit was in the centre of European politics. But he remained humble and continued to live an ascetic life in his simple cell. Francis of Paola died in 1507 and was canonised in 1519.

Saint Francis of Paola is the founder of the Order of Minims, the name reflecting their humility and the wish of being recognised as the least in God's people. Francis himself was a vegan: he abstained from meat and other animal products like eggs and dairy. This can partly be attributed to his desire for religious self-discipline, but partly to his strong belief in non-violence – to humans and animals alike. When Francis established the discipline of his order, he included vegan diet among the rules.

The figure of Saint Francis of Paola is connected with plenty of legends. One of the best known of these is the story about how he crossed the strait of Messina, between Italy and Sicily, when he was refused to board a boat. Francis laid his cloak on the water and used it as a boat. That is why he is the patron saint of sailors.

Some of the legends bear witness to the unusually strong love Francis felt for animals. He is said to have repeatedly rescued animals from hunters, which is a recurring pattern in legends about saints. Another story tells us that, when he was offered a couple of cooked fish by someone who was not aware of Francis' vegan lifestyle, the saint revived the fish and carefully placed them back into the water.

There is a unique legend about Saint Francis of Paola and his pet lamb named Martinello. On one occasion, when Francis was not present, some men who were working on the monastery, slaughtered, roasted, and ate the saint's beloved Martinello, then threw the bones into the furnace. When Francis arrived back, he called the lamb who, lo and behold, walked out of the fire unharmed.

Francis had an even more unusual friendship – with a trout whom he called Antonella. Whenever Francis approached the pool Antonella was living in, the fish came near the shore so that Francis could stroke her. Unfortunately, someone caught and fried the poor animal. When Francis learned what had happened, he revived the fish with the words 'In the name of the Lord, receive your life back again.'[1] Antonella continued to live happily and, so the legend goes, died the same day as Francis.

---

[1] Quoted in Anonymous (1906): The Church and Kindness to Animals. Burns and Oates, London, p. 126.

Christian Saints and Compassion for Animals

There are numerous saints who showed particular kindness to animals and thus should be considered pioneers of animal advocacy.

Saint Blaise, a martyr from the early Christian period, was living in a cave and befriended the animals of the forest. Legend has it that he was visited by herds of wild animals whom he blessed and healed.

Saint Clare of Assisi, a contemporary and follower of Saint Francis of Assisi, was a vegetarian. Moreover, she cared for injured animals, releasing them back into the wild after they had regained health and strength.

Saint Philip Neri, who became known as the Apostle of Rome, was a vegetarian because he did not want animals to be killed. He strongly disapproved cruelty to animals. Philip is said to have released captured mice in distant locations so that they would not disturb people and would not be killed. If flies flew into a house, Philip let them out of the window rather than swatting them.

Worth Reading

Roberts, H. (2004). *Vegetarian Christian saints: Mystics, ascetics and monks*. Sequim: Anjeli Press.

Simi, G. J., & Segreti, M. M. (1977). *St. Francis of Paola: God's miracle worker supreme*. Rockford: Tan Books and Publishers.

Animals Are Back – The Ethics of Jeremy Bentham

For thousands of years, human superiority over animals was taken for granted in all major societies where agriculture had been developed. Animals were regarded as existing solely for the benefit of man. Buddha, Plutarch, and Saint Francis of Assisi were among a handful of thinkers who dared to question unrestricted human dominion over animals. By the Enlightenment, their views had mostly been forgotten. But in the late eighteenth century, the modern animal advocacy movement began to take shape with the appearance of major thinkers who included animals in the moral sphere.

The first notable figure of this period was the French writer and philosopher Voltaire. Voltaire refuted the mechanistic view that animals are mere automata. Based on the erroneous assumption that animals are like machines, unable to feel pain, vivisectionists would nail dogs to tables and cut them open alive, without anaesthesia. Voltaire called these experimenters barbarians. Voltaire's compassion for animals marked the beginning of a new era, the real pioneer of which was Jeremy Bentham.

Born in 1748, Jeremy Bentham was a wunderkind: he learned to read at the precocious age of three and enrolled at Oxford aged 12. The son of a respected attorney, Jeremy Bentham studied law but became disillusioned with the English legal system. After graduation he decided that, instead of practicing law, he would write about it.

Bentham wrote extensively about the inconsistencies of the legal system, and made suggestions on how to improve it. Although Bentham's primary interest was law, he also contributed to the fields of economics and social reforms. Bentham was a true visionary. He supported the abolition of slavery, the equality of women and men, the decriminalisation of homosexuality, and the freedom of the press. Today Jeremy Bentham is remembered as the founding father of utilitarianism. His ideas played a prime role in the history of animal advocacy.

In his 1789 work *An Introduction to the Principles of Morals and Legislation*, Bentham argued that the morality of an act depends on its consequences: an act is moral if it increases happiness, and immoral if it increases suffering. More often

© Springer Nature Switzerland AG 2019

L. Erdős, *Green Heroes*, https://doi.org/10.1007/978-3-030-31806-2_5

than not, a single act affects several beings: it increases the happiness of some individuals but results in suffering for others. In these cases, we have to calculate the net result of the action, i.e., whether it increases or decreases general happiness. For the calculation, all individuals affected by the action have to be taken into account. Also, we must consider the intensity and duration of happiness and suffering. For example, if an act brings slight happiness to a few individuals but causes intense suffering for many others, the act is immoral.

When Bentham said that all individuals affected by the action must be taken into account, he really meant all individuals, irrespective of their sex, race, or species. Bentham insisted that all beings capable of feeling pleasure and pain have interests: they want to experience pleasure and avoid pain. Or, to put it another way, they want to be happy and do not want to suffer. Each individual's interests have to be given equal consideration. For example, the happiness of a poor person is as important as the happiness of a rich one, and the suffering of an animal is no less important than the suffering of a human. This is the principle of equal consideration.

Bentham broke the tradition of excluding animals from moral consideration, which had been based on their assumed lack of intelligence. He pointed out that a human infant is less rational than many animals. Therefore, if we regard human infants as morally considerable, which we certainly should, then we cannot exclude animals from the moral realm either. Bentham concluded that the capacity to reason is morally irrelevant. What is morally relevant is an individual's capacity to feel pleasure and pain. Consequently, no sentient being can be denied equal moral consideration. In a famous and much-quoted passage, Bentham argued this way:

'The day *may* come, when the rest of the animal creation may acquire those rights which never could have been withholden from them but by the hand of tyranny. The French have already discovered that the blackness of the skin is no reason why a human being should be abandoned without redress to the caprice of a tormentor. It may come one day to be recognized, that the number of the legs, the villosity of the skin, or the termination of the *os sacrum*, are reasons equally insufficient for abandoning a sensitive being to the same fate. What else is it that should trace the insuperable line? Is it the faculty of reason, or, perhaps, the faculty of discourse? But a full-grown horse or dog is beyond comparison a more rational, as well as a more conversable animal, than an infant of a day, or a week, or even a month, old. But suppose the case were otherwise, what would it avail? the question is not, Can they *reason*? nor, Can they *talk*? but, Can they *suffer*?' [1]

The weakness of utilitarianism is that happiness and suffering are hard to measure. Thus, it is not easy to assess the expected consequences of a given action. Nevertheless, in many cases reasonable estimates can be made. Take, for example, a new cosmetics product, say, a shampoo. The cosmetics company has two options: it can develop the shampoo using animal tests, or it can rely on cruelty-free methods. Let's assume that the second option is a bit more expensive. Following

[1] Bentham, J. (1879): An Introduction to the Principles of Morals and Legislation. Clarendon Press, Oxford, p. 311.

Bentham's idea, we have to assess the net effects of the two options by taking into account every sentient being who will be affected. If the company uses animals for the testing, the reduced costs may make the owners of the company a bit happier. The product will be cheap, meaning that the consumers will also be slightly happier than they would if the shampoo was more expensive. However, the test animals have to endure unspeakable sufferings. If the company opts for the cruelty-free alternative, the owners of the company have to invest more, which may reduce their happiness. Consumers have to pay a bit more for the shampoo, resulting in a slight decrease in their happiness. But several animals will be spared the torture of the tests. In addition, consumers who are concerned about animals will be happy to buy the product, even if it costs a bit more. Even though it would be hard to make an accurate calculation in this case, and there are several confounding factors, Bentham's guide proves useful. It is certain that, as a result of the huge suffering of the test animals, the net effect of the first option will be far worse than that of the second one. Thus, developing the shampoo with animal tests (or buying the shampoo that was tested on animals) is immoral.

Bentham died in 1832. He was a genius, but a quite eccentric one indeed. His will contained accurate instructions on how his body should be preserved. Bentham's remains, dressed in suits, fitted with a wax head, seated on a chair, are on display in a glass-fronted cabinet at University College London. The body is occasionally brought to the council meetings, where the minutes record Jeremy Bentham as 'present but not voting.'

Jeremy Bentham revived compassion for animals. He was among the first in modern times to recognise that the suffering of animals cannot be ignored. The principle of equal consideration was a huge step in the history of animal protection. The last quarter of the twentieth century saw the renaissance of Bentham's ideas on animals: his views were taken up by Peter Singer, boosting the animal advocacy movement. At long last animals were back in philosophical dialogue and public discourse.

Worth Reading

Bentham, J. (2018). *An introduction to the principles of morals and legislation*. Whithorn: Anodos Books.

Worth Browsing

www.ucl.ac.uk/bentham-project

Pioneers of Animal Advocacy in the Nineteenth and the Early Twentieth Centuries

The nineteenth century witnessed an increasing compassion for animals in England, particularly among the Protestant clergy. The animal advocacy movement gained popular support, spread to other countries and eventually took root in virtually every corner of the world. This chapter is dedicated to some notable characters of the era.

Arthur Broome was one of the greatest pioneers of animal protection. He was born in 1779, graduated from Oxford University, and was ordained an Anglican priest in 1803. Broome thought that protecting animals from wanton cruelty is the obligation of every Christian. Around 1820 he began to plan an organisation that would work on the behalf of animals. On June the 16th, 1824, at the invitation of Broome, a group of like-minded individuals met at Old Slaughter's Coffee House in London. The event proved to be a historic one, as it marked the creation of the Society for the Prevention of Cruelty to Animals, the world's first animal advocacy organisation.

As the secretary of the Society, Broom oversaw the publication of anti-cruelty books and encouraged priests to deliver sermons promoting kindness to animals. The Society hired inspectors who visited markets and slaughterhouses and patrolled the streets of London, investigating animal cruelty. As a result, the Society brought to court hundreds of cases during the first few years alone. The activity was not without dangers: an inspector named James Piper was killed by an angry mob of cockfighters.

By 1826 the organisation had got into financial trouble. As the secretary, Broome had to go to jail and was released only after the other members had managed to collect the sum needed to pay the debt. The bankruptcy and the imprisonment disrupted Broome's life. In 1837 he died in poverty and was buried in an unmarked grave.

As for the Society, the situation began to improve in the years that followed. In 1840 the organisation was granted royal patronage and thus was renamed the Royal Society for the Prevention of Cruelty to Animals (RSPCA). With its history nearing 200 years, today the RSPCA is one of the largest and most respected charities in the UK.

© Springer Nature Switzerland AG 2019
L. Erdős, *Green Heroes*, https://doi.org/10.1007/978-3-030-31806-2_6

Among the founding members of the Society for the Prevention of Cruelty to Animals was Lewis Gompertz. He was an inventor, a reformer who opposed slavery and the oppression of women, and a devoted animal advocate. Gompertz was so passionate about animals that he was a vegan and would never travel by horse-drawn coaches. In 1824 he published a revolutionary book entitled *Moral Inquiries on the Situation of Man and of Brutes*, in which he argued against any use of animals and, decades before Darwin's influential work, he emphasised that humans are not separate from non-human animals.

In the second half of the nineteenth century the emerging animal advocacy movement spread to the US. This process was the result of the activism of early champions, the most important of whom was Henry Bergh.

Henry Bergh was born to a wealthy New York family in 1813. He enrolled at Columbia College but did not finish his studies. Instead, he attended parties, went to the theatre, and generally enjoyed a luxurious lifestyle. Later he started to travel between Europe and New York without any specific goal. From 1862 he served as a diplomat in St. Petersburg, Russia. It was during this time that he became sensitised to the suffering of animals, especially draft animals, who, besides being seriously overloaded and overworked, frequently had to endure merciless beatings. On one occasion Bergh spotted a driver whipping an exhausted donkey. He ordered him to stop and the man obeyed. At that moment Bergh realised how efficient direct intervention on behalf of animals can be.

Bergh spent some time in London, where he met the president of the Royal Society for the Prevention of Cruelty to Animals, which proved to have a lasting impact on Bergh's career. Upon his return to New York, he set up the American Society for the Prevention of Cruelty to Animals (ASPCA) in 1866. Thanks to Bergh's lobbying efforts and good connections, the newly formed organisation was authorised to enforce animal protection regulations. Bergh, who served as the president of the Society, patrolled the streets of New York. Keeping a copy of the animal welfare law in his pocket, he would warn animal abusers that they were breaking the law and, if necessary, arrested the perpetrators. The tall, elegant gentleman wearing a top hat and carrying a cane while keeping a close eye on carriage drivers became the city's trademark. He inspected slaughterhouses and visited markets. On one occasion he arrived at a dogfight like Batman, jumping through the skylight into the pit, to stop the event. In 1867, the ASPCA started an ambulance service for horses and created a network of fountains in the city to provide carriage horses with drinking water. Henry Bergh strongly opposed live pigeon shoots and deserves credit for bringing the use of clay pigeons into fashion.

As a great pioneer Bergh faced significant resistance initially: he received death threats and was ridiculed in the press, usually being referred to as 'the Great Meddler.' But as time went on his single-mindedness gained him respect and by the 1880s he was widely applauded for his work.

When Bergh was asked why he was spending so much time helping animals instead of alleviating human suffering, he responded that if animals had to wait until all injustices among humans were solved, they would still be waiting at the Second Coming. But by no means was Bergh indifferent to human suffering. In 1874 he was informed about a young girl who had been brutally abused by her foster mother. Bergh intervened promptly, rescuing the child and bringing the foster mother to court. The next year he founded The New York Society for the Prevention of Cruelty to Children. 'Mercy to animals means mercy to mankind,'[1] Bergh said.

Henry Bergh and the ASPCA inspired many Americans to take action and form similar organisations. Caroline Earle White was among Bergh's earliest followers.

Caroline Earle White was born in Philadelphia, Pennsylvania in 1833. Her parents supported African Americans and heavily criticised slavery. Caroline became an acclaimed writer: she published successful short stories and novels. Besides her literary work, she helped orphaned children and the poor, and was also involved in the women's rights movement. But she is best remembered for her animal advocacy activism.

After Henry Bergh founded the ASPCA, Caroline Earle White asked him for advice on how to launch a sister organisation in Philadelphia. In 1867 White set up the Pennsylvania Society for the Prevention of Cruelty to Animals (PSPCA). However, at that time women were not allowed to lead organisations. As White was excluded from the management of the organisation she had co-founded, she created the Women's Branch of the PSPCA.

The Women's Branch started the first animal shelter in the US. Also, the organisation employed inspectors to prevent and investigate animal cruelty cases. In addition, they successfully lobbied for legislative improvements, campaigned against fox hunting, pigeon shooting, dogfights, cockfights, and started a humane education programme. Today the organisation operates under the name Women's Animal Center.

When the use of animals in research was gaining ground in the US, Caroline opened a new front. In London she met Frances Power Cobbe, the leader of the anti-vivisection efforts in the UK. Back in America, Caroline set up the American Anti-Vivisection Society (AAVS) in 1883. The long-term goal of the organisation is the abolition of all animal experiments, while the short-term objective includes reducing the number of animals used in tests, ending classroom dissections, and achieving improved standards for laboratory animals. In 1892 the Society launched the *Journal of Zoophily*, which was edited by Caroline Earle White and Mary Frances Lovell.

By the end of the nineteenth century the animal protection movement had gained considerable momentum in several countries. However, while there was an increasing number of on-the-ground activists, the philosophical discussion about the rights of animals was only a side issue. This changed when Henry Stephens Salt arrived on the scene.

Henry Stephens Salt was born in Naini Tal, India in 1851, but spent his life in England from 1852. After graduating from the University of Cambridge, Salt taught classics at Eton College. Influenced by the thoughts of Plutarch, Henry David Thoreau, and Percy Bysshe Shelley, Salt became interested in living a simple and compassionate life. Thus he adopted vegetarianism, left his job, and moved to a small rural cottage with his wife.

---

[1] Quoted in Furstinger, N. (2016): Mercy: The Incredible Story of Henry Bergh, Founder of the ASPCA and Friend to Animals. Houghton Mifflin Harcourt, Boston, p. vi.

A prolific writer, Salt published dozens of books, including biographies and writings on various topics such as nature, poetry, and, most importantly, animal advocacy. His first book, *A Plea for Vegetarianism* was published in 1886.

Salt's best-known work, *Animals' Rights Considered in Relation to Social Progress*, is now regarded as a classic. As the first comprehensive philosophical defence of animals' rights, the book has a historic importance, while its content is as relevant today as it was when it was first published at the end of the nineteenth century. Embracing the ethical implications of Darwinian evolution, Henry Salt refused the existence of a gulf between humans and non-human animals. Viewing human rights and animal rights as joint issues, Salt was convinced that all sentient beings, human and non-human, have rights. 'If "rights" exist at all – and both feeling and usage indubitably prove that they do exist – they cannot be consistently awarded to men and denied to animals, since the same sense of justice and compassion apply in both cases,'[2] Salt argued. Consequently, he considered hunting, shooting, vivisection, and the killing of animals for food or clothing morally unacceptable. As Salt put it in one of his later works, *The Story of My Cousins*, '…when the oneness of life shall be recognized, such practices as blood-sports will be not only childish but impossible; vivisection unthinkable; and the butchery of our fellow-animals for food an outgrown absurdity of the past.'[3]

Worth Reading

Beers, D. L. (2006). *For the prevention of cruelty: The history and legacy of animal rights activism in the United States.* Athens: Swallow Press/Ohio University Press.

Furstinger, N. (2016). *Mercy: The incredible story of Henry Bergh, founder of the ASPCA and friend to animals.* Boston: Houghton Mifflin Harcourt.

Phelps, N. (2007). *The longest struggle: Animal advocacy from Pythagoras to PETA.* New York: Lantern Books.

Salt, H. S. (2009). *Animals' rights considered in relation to social progress.* Ithaca: Cornell University Library.

Worth Browsing

www.rspca.org.uk
www.aspca.org
www.pspca.org
womensanimalcenter.org
aavs.org

---

[2] Salt, H. S. (1894): Animals' Rights Considered in Relation to Social Progress. MacMillan & Co, New York, p. 19.

[3] Quoted in Stroup, W.: Henry Salt on Shelley: Literary Criticism and Ecological Identity. https://romantic-circles.org/praxis/ecology/stroup/stroup.html

Charles Darwin – Evolution and the Fall of Human Superiority

Charles Darwin was, without any doubt, one of the most brilliant minds in the history of science. But Darwin was more than a genius: he was also a revolutionary who redefined humanity's place in the universe. The ethical implications of Darwinian evolution call for a re-evaluation of how *Homo sapiens* should treat non-human animals.

Charles Darwin was born in 1809 in Shrewsbury, England. In his youth he showed vivid interest in plants and animals. He was also keen on chemistry, carrying out experiments with his brother in the garden shed. Aged 16, in the hope that he would follow in his father's and grandfather's footsteps of becoming a doctor, Charles was sent to Edinburgh to study medicine. But he did not like the idea, became disinterested, and quit the university. In 1827 he started his studies in Cambridge to become a clergyman. Although Darwin would never work as a priest, the years spent in Cambridge had a profound influence on his career. As was usual at the time for those interested in theology, Darwin continued to study natural sciences. He met and befriended John Henslow, a well-known professor of geology and botany. Darwin often went on tours with the professor and thus became known as 'the man who walks with Henslow.'

Just after Darwin's graduation in 1831, the British navy ship Beagle was about to set sail to circumnavigate the Earth, and the captain was seeking a naturalist to join the crew. Henslow recommended Darwin. Darwin was happy to grab the chance but his father strongly opposed the adventure, declaring that he would give his consent only if Charles could find 'any man of common sense' who regarded the voyage as a good idea. It soon turned out that Charles' uncle considered the expedition an opportunity not to be missed. As the uncle undeniably qualified as a man of common sense, the young Charles Darwin could accept the position.

In December 1831 the Beagle left England and crossed the Atlantic Ocean to reach South America. While the rest of the crew was surveying the coasts, Darwin was undertaking inland expeditions, observing the geology and wildlife of the continent. In the autumn of 1835 the Beagle spent 5 weeks in the Galápagos Islands, where Darwin observed the now famous giant tortoises and finches. The Beagle

© Springer Nature Switzerland AG 2019
L. Erdős, *Green Heroes*, https://doi.org/10.1007/978-3-030-31806-2_7

then sailed to Australia, visited Cape Town, returned to South America, and finally landed in England in October 1836.

When Darwin boarded the Beagle in 1831, he, like almost everyone else in the western world, believed that species were fixed and immutable. By the time he arrived back in England in 1836, he was considering the possibility of evolution, with species gradually changing over time and turning into new species. The task now was to discover how evolution works. Darwin had to find the mechanism underlying evolutionary changes.

The young scientist started his investigation and found the answer in a surprisingly short time. Darwin's explanation starts with the well-known observation that living beings produce more offspring than can survive to adulthood. Offsprings tend to resemble their parents, i.e., they inherit their characteristics from their ancestors. However, individuals within a species are never exactly the same, that is, there are smaller or larger differences between them. As more offspring is produced than can survive, there will be a competition for resources such as food, nesting places, and the like. Also, organisms have to escape predators and cope with harsh environmental conditions. The fact that there is variation among the individuals entails that some of them will have better chances to survive than the others. For example, better eyesight enables the earlier detection of enemies or prey, thicker fur increases survival chances during the winter, and a slightly longer neck or finger or tongue makes new resources available. Individuals with advantageous characteristics are more likely to survive and breed. Their offspring will inherit the advantageous characteristics. This means that in the next generations individuals with favourable characteristics will be more and more numerous, while individuals that are less well adapted to the environment will gradually disappear. Over countless generations the small modifications add up, resulting in completely new species. This is how variation and natural selection drives evolution.

Darwin managed to solve the greatest riddle in biology but he delayed publishing the groundbreaking discovery for two decades. He wanted to gather evidence to prove his theory beyond any argument. He carried out experiments on plants, corresponded with scientists and breeders, and carefully examined the scientific literature. Another reason for the delay was Darwin's fear that his work would cause public outrage, threatening the religious worldview of the time.

In 1842 Darwin and his family moved to a small village not far from London. By this time he was already a respected scientist due to his contributions about issues not related to the evolutionary theory. While Darwin was quietly conducting his scientific inquiry into evolution, he was suffering from a mysterious illness. Attempts to cure the disease failed and the cause of the malady is still unknown. It is possible that he had contracted a tropical infection during the voyage of the Beagle. Some experts speculate that his symptoms may have been caused by the stress about how his discovery would be received by the religious society of the time. Considering the idea that species can change 'is like confessing a murder,'[1] Darwin wrote to one of his colleagues.

---

[1] Darwin's letter to Joseph D. Hooker. http://www.lettersofnote.com/2009/09/it-is-like-confessing-murder.html

In 1858 Darwin could not postpone publishing his results any longer. In June of that year, naturalist Alfred Russel Wallace sent Darwin a manuscript outlining the theory of evolution by natural selection. Darwin and Wallace had come to the very same conclusion independently. Darwin's friend, the distinguished geologist Charles Lyell solved the problem of priority. The papers of both Darwin and Wallace were published in the same issue of the Linnean Society's journal. It has to be emphasised here that Wallace never doubted Darwin's priority and acknowledged that Darwin had accumulated a much larger body of evidence supporting the theory.

In 1859 Darwin published *On the Origin of Species*. All copies of the book were sold out on the first day. The reception, as Darwin had expected, was not entirely positive. While many readers were ready to accept Darwin's claims, others wanted to defend the old worldview, according to which species cannot change. Though Darwin had carefully avoided referring to the origin of the human species in his work, it was evident that, if he was right, the principle applied to our species as well, meaning that humans had descended from non-human animals. For many people in the nineteenth century this was simply unacceptable. For them, questioning the privileged status of humans amounted to a scandal.

A heated debate unfolded, but Darwin's argument was so strong, supported by so much evidence that it became generally accepted within a relatively short period. Everything that has been revealed in biology since 1859, including the discoveries of modern cell biology, genetics, and molecular biology, has not only verified Darwinian evolution, but made it even more robust. Those few who still doubt evolution simply want to preserve the illusion of human uniqueness and superiority. Their 'arguments' are nothing more than pseudo-science, akin to alchemy and the flat-earth theory.

Darwin continued his work in several branches of biology and published many important books. In *The Descent of Man, and Selection in Relation to Sex* (1871) and *The Expression of the Emotions in Man and Animals* (1872) Darwin shows how closely humans are related to other animals, not only in physical characteristics, but also regarding mental faculties and emotions.

Darwin, probably the greatest biologist of all times, died in 1882. He was laid to rest in Westminster Abbey.

Are Chimpanzees Humans?
It is well known that chimpanzees and bonobos (also called pygmy chimpanzees) are our closest living relatives. So much so that, according to some researchers, chimps and bonobos should be regarded as humans.

Chimpanzees and bonobos have commonly been classified into the genus *Pan* (chimpanzees). Our species, *Homo sapiens*, and other, extinct human species belong to the genus *Homo* (humans). However, both fossil- and DNA-based evidence suggest the relation is so close that it would be reasonable to include chimpanzees, bonobos, and humans in the very same genus. Thus, the genus *Homo* should include chimpanzees and bonobos, which would mean

(continued)

accepting chimps and pygmy chimps as humans. This would have far-reaching ethical implications. To experiment on humans is not acceptable, even if they happen to be a bit hairier than most of us are. Also, the eradication of our forest-dwelling African cousins is a crime against humanity.

Treating chimpanzees and bonobos as humans may well be justified from a scientific perspective, but it is resisted by those who want to see *Homo sapiens* as the pinnacle of creation. However, sooner or later reality will triumph over myth.

There were some scientists before Darwin who suggested that species can change and evolution does happen. But Darwin was the first who could figure out how evolution operates. Moreover, he was able to accumulate overwhelming evidence in favour of it. The scientific impact of his work cannot be overestimated. But the ethical implications of Darwinian evolution are equally important.

First, it became evident that evolution is driven by natural forces, which applies to the human species as well. Thus, *Homo sapiens* has evolved just as all other species have. We are not separate from nature, nor are we superior to the millions of other species.

Second, there is no chasm between humans and non-human species. As Darwin usually emphasised, there is no fundamental difference between humans and our closest relatives, apes and monkeys. Darwin did not deny that differences do exist, but these are differences of degree, not of kind. All human characteristics have evolved gradually, and are present – though sometimes only in a rudimentary form – in several other animals as well.

Third, as humans are not radically different from several other animals, there is no morally acceptable reason to treat animals differently simply because they happen to belong to other species.

Fourth, in the light of Darwinian evolution, to regard *Homo sapiens* as 'the most developed' or 'the most advanced' species is a nonsense. All existing species are equally advanced, because all have equally long evolutionary histories. Consequently, there are no 'higher' and 'lower' animals. 'It is absurd to talk of one animal being higher than another,'[2] Darwin wrote. The idea of the great chain of being, with humans at the top, has to be abandoned once and for all.

Fifth, Darwin guessed correctly that all living beings on Earth descended from a common ancestor. Literally, all living creatures are our relatives.

Sixth, Darwin realised that species are abstract categories, created by the human mind, but not existing in nature. As Darwin himself put it, 'I look at the term species, as one arbitrarily given for the sake of convenience to a set of individuals closely resembling each other, and that it does not essentially differ from the term

---

[2] Charles Darwin's words in his notebook, quoted in: The Annotated Origin: A Facsimile of the First Edition of On the Origin of Species. By Charles Darwin, annotated by James T. Costa. The Belknap Press of Harvard Universigty Press, Cambridge, p. 441.

variety, which is given to less distinct and more fluctuating forms. The term variety, again, in comparison with mere individual differences, is also applied arbitrarily, and for mere convenience sake.'[3] The species concept is useful in everyday speech and probably indispensable in science. But we always have to keep in mind that species as such are not real.

To sum it up, the almost religious awe surrounding the human species is not justified. Anthropocentric thinking was shattered by Nicolaus Copernicus, Giordano Bruno, and Galileo Galilei. Charles Darwin finished the job and smashed the remainder of human-centred views. More than 150 years after the publication of Darwin's historic work, it is time to consign anthropocentrism to the dustbin of history.

Worth Reading

Darwin, C. (2004). *The descent of man and selection in relation to sex*. London: Penguin Books.
Darwin, C. (2009a). *On the origin of species by means of natural selection*. London: Penguin Books.
Darwin, C. (2009b). *The expression of the emotions in man and animals*. London: Penguin Books.
Rachels, J. (1990). *Created from animals: The moral implications of Darwinism*. Oxford: Oxford University Press.

Worth Browsing

darwin-online.org.uk

Worth Watching

Darwin's Brave New World (2009)
Charles Darwin and the Tree of Life (2009)

---

[3] Darwin, C. (1869): On the Origin of Species by Means of Natural Selection. John Murray, London, pp. 62–63.

Reverence for Life – Albert Schweitzer's Biocentrism

'If you must be human, then be Albert Schweitzer,'[1] Czech philosopher Erazim Kohák wrote. Schweitzer was surely among the greatest personalities of the twentieth century. He became a legend in his lifetime and continues to inspire millions all over the world, while his philosophy may be regarded as a much-needed moral compass in the twenty-first century.

Albert Schweitzer was born in 1875 in Kaysersberg, Germany (now France), and was brought up in Günsbach, where his father served as a Lutheran pastor. Albert Schweitzer's empathy for all living beings became apparent at a very young age. As a kid he developed the practice of reciting his own prayer each evening: 'Dear God, protect and bless all living beings. Keep them from evil and let them sleep in peace.'[2] A decisive experience of his childhood happened when he was nine. One of his friends suggested that they should go hunting for birds, using catapults. Albert knew this was wrong, but, afraid of being ridiculed, he did not dare to refuse. As they took aim, the bell of the nearby church began to ring. To Albert, this was a warning not to commit the sin. He shooed the birds off and ran away.

Schweitzer learned to play the organ as a child. In 1893 he started to study theology and philosophy in Strasbourg. He obtained the doctoral degree in philosophy in 1899 and was awarded the licentiate in theology one year later. In 1902 he became a lecturer in Strasbourg.

In 1896 Schweitzer decided that, from age 30, he would devote his life to serving humanity directly, although at that time he did not know exactly what this service was to be. In 1904 he came across an advertisement seeking medical personnel for a mission in French Equatorial Africa (now Gabon). Schweitzer felt he had found his calling. Thus, to the greatest astonishment of his family and colleagues, he gave up his career as an organist, musicologist, philosopher, theologian, and university

---

[1] Kohák, E. (1998): Varieties of Ecological Experience. In: Cohen, R. S.; Tauber, A. I. (eds.) Philosophies of Nature: The Human Dimension, Springer, pp. 257–271.

[2] Quoted in Martin, W.; Ott, M. (2017) (eds.) Albert Schweitzer's Reverence for Life: The Adventure of Being True to Yourself. Lulu Self-Publishing, p. 19.

© Springer Nature Switzerland AG 2019
L. Erdős, *Green Heroes*, https://doi.org/10.1007/978-3-030-31806-2_8

lecturer, and started to study medicine. After finishing his medical studies he attended additional courses in tropical medicine. In 1913 Schweitzer and his wife, Hélène Bresslau set off for Africa. They settled in Lambaréné, on the shore of the Ogowe River, at the edge of the jungle.

The Schweitzers began their medical work in the open air, then they refitted an old henhouse so it could serve as a clinic until the new hospital building was ready. They treated thousands of patients during the first year alone. Schweitzer thought their work could be understood as a first step in compensating African people for centuries of colonialism.

In 1917, German citizens in French territory, Schweitzer and his wife were arrested and transported to France as prisoners of war. After being released in 1918, he worked as a medical assistant and an assistant pastor. After fund-raising and lecture tours, he returned to Africa in 1924. With the help of the patients and their families, Schweitzer began to rebuild and enlarge the hospital, which had deteriorated during his absence.

Schweitzer's work in the jungle captured public imagination. By the 1920s he had become famous and respected all around the world. In the next decades he was frequently invited to give organ concerts and talks in various European and American cities, and received several awards and honorary doctorates.

In 1953 Albert Schweitzer was awarded the Nobel Peace Prize. But even at the peak of his fame he retained his humble demeanour. When he was asked why he travelled in third-class carriages he responded that he chose the third class only because there was no fourth class.

Towards the end of his life, besides working at his hospital, he became active in the anti-nuclear movement. He strongly opposed nuclear weapons tests and urged politicians to end the atomic arms race. His radio speeches about the dangers of nuclear tests were published under the title *Peace or Atomic War*. Schweitzer died in 1965, at the hospital he had founded more than five decades earlier.

Albert Schweitzer was a multi-talent: a musician, scholar, churchman, doctor, and peace activist at the same time. His humanitarian work earned him worldwide admiration. But according to Schweitzer himself, his most important achievement was the elaboration of a new ethic that can be summarised by the phrase 'reverence for life.'

Schweitzer firmly believed that ethical systems that restrict their attention to humans are incomplete. He was pursuing an ethical principle that includes all living beings. On a September day in 1915, he was travelling upstream on the Ogowe River. At sunset, as the boat was moving past a herd of hippos, Schweitzer was suddenly struck by the idea of 'reverence for life,' which he developed into a biocentric (i.e., life-centred) ethical system.

Schweitzer's argument starts with a very simple fact: as a human being I can easily recognise that I live and have an interest in going on living. I also have to acknowledge that every single living creature has the very same interest. Indeed, the interest in continued life is common to all living beings. Granted, many living organisms are not conscious. However, consciousness is morally irrelevant here, as even those who are not aware of their interests do have an interest in continued existence. I rightfully expect that my life and my interest in continued existence be respected. And fairness requires that I treat the interests of every other living being with the same respect. Schweitzer summarised his ethics this way: 'It is good to maintain and encourage life; it is bad to destroy life or to obstruct it.'[3]

According to Schweitzer, all living beings are fundamentally equal. 'The ethics of reverence for life makes no distinction between higher and lower, more precious and less precious lives,'[4] he stated. Of course, Schweitzer was aware that harming other living beings is unavoidable in some cases; for example, we have to kill bacteria to protect ourselves. However, Schweitzer insisted that all such cases are based on subjective decisions and thus cannot form the basis of a moral hierarchy. He added that harming other living beings is acceptable only if this is absolutely necessary.

Schweitzer lived out his ethic, making his life his argument. He devoted his life to helping people who had poor access to medical care. But he also cared for injured or orphaned animals, some of whom remained permanently in the proximity of the hospital, turning the facility into a kind of sanctuary. Sizi, a female cat rescued by Schweitzer, was his companion for 23 years. Schweitzer cared for monkeys, apes, antelopes, and birds, and he would reportedly put his pet wild pigs to bed with Brahms' Lullaby. One day Schweitzer adopted three orphaned pelicans. Once full-grown, two of them flew away, but the third one, named Parsifal, remained with Schweitzer, standing on guard in front of the doctor's door every night. *The Story of My Pelican* is an account of this unusual friendship.

But Schweitzer took the practice of reverence for life further. All of his everyday actions were guided by his ethic. He carefully captured and freed insects that accidentally flew indoors in the daytime. At night he would keep his windows closed while working by lamplight in the hot equatorial climate so that insects could not enter the room where they would have been killed by the kerosene lamp. During construction or reconstruction works on the hospital, he took the greatest care to minimise the effects on the environment.

'Ethics is responsibility without limit towards all that lives,'[5] Schweitzer wrote. He knew that to show respect to all living beings is at times hard, as it used to be hard to show compassion for slaves. But, in Schweitzer's words, 'Constant kindness can accomplish much. As the sun makes ice melt, kindness causes misunderstanding, mistrust, and hostility to evaporate.'[6]

---

[3] Schweitzer, A. (1955): Civilization and Ethics. Adam & Charles Black, London, p. 242.

[4] Schweitzer, A. (1966): The Teaching of Reverence for Life. Peter Owen, London, p. 47.

[5] Schweitzer, A. (1955): Civilization and Ethics. Adam & Charles Black, London, p. 244.

[6] Quoted in Gibb, Y.; Gibb, A.; Bennett, D. (2018): Kindness and the Independent Owner Managed Business. In Haskins, G.; Thomas, M.; Johri, L. (eds.) Kindness in Leadership. Routledge, Abingdon.

Taking Action for Animals

All of us have plenty of ways to help animals every day. If you buy cosmetics or household products that have been tested on animals, you support cruel and unnecessary animal experiments. It only takes a little time to find cruelty-free alternatives. Pay attention to the labels on the products or find cruelty-free companies and brands on the internet!

Consider becoming vegan or vegetarian! If you regard this as too demanding, reduce your meat consumption. Only buy meat, eggs, and dairy that comes from free-ranging animals. Do your best to avoid products from factory farms, as these facilities are extremely cruel (in addition, their products are usually unhealthy). In many countries (including all member states of the European Union) egg labelling is compulsory, so you can easily find out how the chickens are being kept. If you purchase seafood products, prefer those with the Dolphin Safe or the MSC (Marine Stewardship Council) label.

Support on-line petitions calling for animal protection laws and urging companies to show compassion for animals.

Do not wear fur and never buy trophies, ivory products or souvenirs made of animals.

If you can, support animal advocacy groups. Volunteering for animal shelters is a useful and enjoyable pastime; both the staff and the animals will be grateful to you. Adopt stray animals instead of buying from breeders. Spay or neuter your pets to avoid overpopulation.

Avoid visiting roadside zoos and marine parks, do not attend bullfights or similar events.

Reduce your ecological footprint: a smaller environmental impact means increased chances for wild animals.

Schweitzer's deep Christian faith was in consonance with his biocentric outlook. As a minister of the Lutheran church he regularly preached on Sundays, and reverence for life was a recurring topic in his sermons. For him it was clear that the commandment 'You shall not kill' (Exodus 20:13) prohibits taking the lives of humans and non-humans alike. As Schweitzer put it, 'Reverence concerning all life is the greatest commandment in its most elementary form.'[7] In the New Testament Jesus says 'Truly I tell you, whatever you did for one of the least of these brothers and sisters of mine, you did for me.' (Matthew 25:40) Schweitzer had no doubt that even the smallest creatures are our brothers and sisters, thus helping them is a moral obligation. Albert Schweitzer the believer and theologian did not hesitate to reject human-centred views prevalent among many Christian thinkers.

Schweitzer opposed sport hunting, the use of animals for entertainment, and factory farming, and eventually became a vegetarian. Of animal experiments he wrote,

---

[7]Albert Schweitzer's sermon (February the 16th, 1919), translated by Reginald H. Fuller. Schweitzer, A. (1993): Reverence for Life. Irvington Publishers, New York, p. 116.

'Those who experiment upon animals by surgery and drugs, or inoculate them with diseases in order to be able to help mankind by the results obtained, should never quiet their consciences with the conviction that their cruel action may in general have a worthy purpose. In every single instance they must consider whether it is really necessary to demand of an animal this sacrifice for men.'[8]

Albert Schweitzer was a person of exceptional moral integrity: his thoughts and actions were in perfect harmony. Through his life he showed that it is possible to live according to the principle of reverence for life. And his example encourages us to try the same.

Worth Reading

Free, A. C. (Ed.). (1988). *Animals, nature and Albert Schweitzer*. Washington, DC: The Flying Fox Press.
Schweitzer, A. (1965a). *The teaching of reverence for life*. New York: Holt, Rinehart and Winston.
Schweitzer, A. (1965b). *The story of my pelican*. New York: Hawthorn Books.
Schweitzer, A. (1987). *The philosophy of civilization*. New York: Prometheus Books.
Schweitzer, A. (1989). *A place for revelation: Sermons on reverence for life*. New York: Macmillan Publishing.
Schweitzer, A. (2009). *Out of my life and thought: An autobiography*. Baltimore: The Johns Hopkins University Press.

Worth Watching

Albert Schweitzer: My life is my argument (2005)
Albert Schweitzer (1956)

---

[8] Quoted in Free, A. C. (ed.) (1988): Animals, Nature and Albert Schweitzer. The Flying Fox Press, Washington, p. 36.

Cleveland Amory and the Fund for Animals

Kindness to humans is considered a virtue in our societies. But real moral integrity requires kindness to animals as well. Kindness was the key concept in the life and work of Cleveland Amory, the successful writer, famous television personality, and devoted animal advocate.

Born in 1917, Cleveland Amory grew up in Boston. He enrolled at Harvard University, where he was editor of the university's student newspaper. After graduating in 1939, Amory worked as an editor at *The Saturday Evening Post*. Between 1941 and 1943 he served in military intelligence in World War Two. After the war he was a reporter. Sometime in the mid-1940s he had to cover a bullfight. He was disgusted with the event. The shocking experience had a lasting effect on his career as an animal advocate.

In 1947 Amory published *The Proper Bostonians*, a satirical depiction of the social elite in Boston. The book was followed by *The Last Resorts* and *Who Killed Society?*, the bestselling trilogy making Amory a well-known author throughout the US. Amory wrote articles for various publications and also appeared on radio. For 11 years he was a commentator of *The Today Show*, a popular morning television programme, but was fired because he expressed his views about animal advocacy issues, satirically criticising sport hunting and needlessly cruel animal experiments.

Cleveland Amory was convinced that the purpose of human life is to help others, especially the weak. From the 1960s he became increasingly involved in animal advocacy. His articles often discussed topics related to animal protection, and he joined the board of directors of The Humane Society of the United States, the best-known animal advocacy group at the time. But he explored new ways to help animals, establishing The Fund for Animals in 1967. In its initial years the Fund lobbied and petitioned for endangered animals, and recruited celebrities for the anti-fur campaign titled 'Real People Wear Fake Fur.'

In 1978 the Fund's donation enabled Paul Watson to purchase a vessel to interfere with the annual slaughter of baby seals in Canada. Cleveland Amory joined Watson's crew and the team managed to save over a thousand seal pups by spray-painting their pelts with a harmless dye that made them worthless to the sealers.

© Springer Nature Switzerland AG 2019
L. Erdős, *Green Heroes*, https://doi.org/10.1007/978-3-030-31806-2_9

The next large-scale campaign of Cleveland Amory and The Fund for Animals was an action to save the burros of the Grand Canyon National Park. By the mid-twentieth century, runaway and released burros had established a self-sustaining wild population, which seemed to threaten the ecological balance of the area. The National Park Service decided to solve the problem by shooting the animals. However, Cleveland Amory was sure there must be a more humane way of protecting the natural values. He hired cowboys to catch the burros and a Vietnam veteran helicopter pilot to airlift the animals out of the canyon. The operation saved 577 burros without a single injury to burros or humans, showing that it is possible to find a compromise that is acceptable for both conservationists and animal advocates.

Amory managed to find temporary homes for the rescued burros, but it became evident that the Fund needed a permanent shelter. Amory had long been nurturing a plan to set up a sanctuary for abandoned and abused animals. The Fund bought a ranch in Texas and developed it into Black Beauty Ranch – the name was inspired by Anna Sewell's bestselling novel *Black Beauty*.

The first resident of the sanctuary was a cat who arrived half-dead, dragging a leghold trap when the sanctuary was still under construction. She was nursed back to life, though her leg had to be amputated.

One of the most famous residents of the Black Beauty Ranch was Nim Chimpsky, a chimpanzee who was taught to communicate using sign language. Once the linguistic study ended, Nim was sold to a laboratory, where he was kept under horrific conditions. Nim was rescued by Cleveland Amory, brought to the Black Beauty Ranch, where the chimp spent the rest of his life in peace.

Today the facility operates under the name The Cleveland Amory Black Beauty Ranch. It is home to more than 800 animals rescued from slaughterhouses, laboratories, trophy hunting ranches, roadside zoos, and cruel pet owners. The Fund operates the Duchess Sanctuary, a safe haven for horses, donkeys, and ponies, and a wildlife centre that cares for injured and orphaned wild mammals and birds. The Fund's Rural Area Veterinary Services programme provides free veterinary care and spay-neuter surgeries for poor rural inhabitants.

The Fund's largest rescue mission was the removal of thousands of feral goats from San Clemente Island. The military planned to shoot the animals but Amory intervened by asking Secretary of Defense Casper Weinberger, a former Harvard classmate, to grant the Fund permission to relocate the animals.

In the 1990s The Fund for Animals played a major role in stopping the infamous live pigeon shoot of Hegins, Pennsylvania. Organised every Labour Day, the event included the gunning down of over 5000 birds each year, just for fun. Pigeons were kept in small cages for days, typically without food and water. When the shooters were ready, a special apparatus threw the pigeons into the air, only to be shot down by the 'sportsmen.' A small proportion of the birds died instantly, others were collected and killed by local kids, or were wounded and died later. The Fund organised demonstrations. Some animal advocates entered the killing fields and released as many pigeons as was possible before the police arrived. The activists were arrested and fined. Some of them, including Heidi Prescott from the Fund and Ingrid Newkirk from PETA, refused to pay the fine and thus had to go to jail. The Fund for Animals also organised rescue teams, which collected wounded birds and provided them with veterinary care. Moreover, the Fund started to lobby for a ban on live pigeon shoots. The efforts paid off in 1999, when the live pigeon shoot of Hegins was cancelled permanently.

The Fund for Animals succeeded in stopping several other sport hunting, bear wrestling, and cockfighting events and their efforts led to legal restrictions regarding the use of bait, dogs, and certain types of traps for hunting.

Besides his activity as the president of The Fund for Animals, Amory used his literary skills to promote the idea of animal advocacy. In 1974 he published *Man Kind?*, exposing the horrors of hunting and the fur industry. His next book, *Animail*, discussing topics related to animals, came out in 1976.

On Christmas Eve 1977, Amory rescued and adopted a starving stray cat from a New York street. Polar Bear, as the cat was named, became the subject of the book *The Cat Who Came for Christmas*, and two sequels, *The Cat and the Curmudgeon*, and *The Best Cat Ever*. All three parts of the trilogy were instant bestsellers, making Polar Bear the world's most famous cat. Amory's beloved cat is buried at the Black Beauty Ranch.

Cleveland Amory, who passed away in 1998, sometimes referred to animal advocates as soldiers of the Army of the Kind. This army is becoming stronger each day, but it still needs further reinforcement. By helping animals, all of us can join Cleveland Amory's Army of the Kind.

Worth Reading

Amory, C. (1974). *Man kind? Our incredible war on wildlife*. New York: Harper & Row.

Amory, C. (1976). *Animail*. New York: Windmill Books/E. P. Dutton & Co..

Amory, C. (1998). *Ranch of dreams*. Thorndike: Thorndike Press.

Amory, C. (2002a). *The cat and the curmudgeon*. Boston: Little, Brown and Company.

Amory, C. (2002b). *The best cat ever*. Boston: Little, Brown and Company.

Amory, C. (2013). *The cat who came for Christmas*. Boston: Little, Brown and Company.

Marshall, J. H. (2006). *Making burros fly: Cleveland Amory, animal rescue pioneer*. Boulder: Johnson Books.

Worth Browsing

www.fundforanimals.org

Worth Watching

Making Burros Fly (2010)

Spokesmen for Animals – Richard Ryder, Peter Singer, and Tom Regan

Having earned a degree in psychology from the University of Cambridge, Richard D. Ryder was working as a clinical psychologist in Oxford. He had firsthand experience of the horrors of animal experiments. Unlike many of his colleagues, Ryder did not harden his heart and never got accustomed to the animals' suffering. Instead, he went over to the other side and joined the animal advocates. Ryder met a handful of like-minded individuals, with whom he set up a small informal circle, which became known as the Oxford Group. They started to organise demonstrations against bloodsports and the use of animals in research. More importantly, the Oxford Group turned out to be the cradle of ideas that would have a lasting effect on the course of animal advocacy.

It was during his time in Oxford that Ryder coined the term 'speciesism.' Speciesism describes a moral standpoint that is akin to racism. Both views try to justify the mistreatment of others based solely on taxonomic position (race or species). Racism and speciesism treat individuals who are not in the preferred taxonomic group as not belonging to the moral sphere. However, taxonomic position in itself is morally irrelevant. Thus, so Ryder's argument runs, excluding animals from the moral community is unfair, just as excluding certain human races from the moral community is unacceptable.

Ryder has published several influential articles and books. In *Victims of Science,* arguably his most important contribution, Ryder shows that the majority of animal tests are totally needless and will be of no use to society. He also describes how and why young experimenters, who initially feel sorry for the animals, gradually become desensitised.

Ryder also invites his readers to carry out a thought experiment. Suppose we were visited by members of a technically advanced extraterrestrial civilisation, who started to carry out deadly experiments on us, kill us for fun, and breed humans to eat them. If the aliens argued that it was morally acceptable for them to use us, no human would ever accept this. But then why do so many of us accept that we can do all these things to animals?

© Springer Nature Switzerland AG 2019
L. Erdős, *Green Heroes*, https://doi.org/10.1007/978-3-030-31806-2_10

Focusing on animal experiments, Ryder emphasises that experimenters tend to use inconsistent arguments. They claim that the test animals are so *similar* to us that they serve as useful models of humans, for example in the study of diseases. The very same experimenters insist that sacrificing animals is morally permissible because animals are very *different* from us.

Richard Ryder's work has had a crucial impact on animal ethics, activism, and European animal welfare legislation.

Australian philosopher Peter Singer spent some time in Oxford as a student and became a member of the Oxford Group soon after its formation. His interest in animal ethics began when he was having lunch with one of his vegetarian classmates. After a short talk Singer realised that treating animals as mere raw materials is unacceptable. Thus he adopted vegetarianism and started to examine our relation to animals more closely. Today Peter Singer is widely considered the world's most influential living philosopher and a leading figure in current animal activism. His 1975 book *Animal Liberation* re-energised the animal advocacy movement, became a classic, and continues to be a must-read for anyone who loves animals.

Although *Animal Liberation* is a work of philosophy, its style is easy to understand, which has contributed to its unparalleled success. As a starting point, Singer argues that the principle of equality applies to all humans, irrespective of their race, sex, intelligence, or any similar characteristic. Granted, there are countless differences among the different races, between women and men, and among all human individuals. But all these differences are morally irrelevant: whatever someone's race, sex, physical or mental capacities are, their interests cannot be ignored. Everyone's interests have to be given equal consideration. Following Jeremy Bentham, Singer adds that the principle of equality applies to non-human animals as well. If, among humans, the colour of the skin or the level of intelligence does not entitle someone to belittle the interests of others, how could any similar characteristic entitle humans to belittle the interests of animals? What is relevant is the capacity for suffering. We cannot mistreat other humans because we have to avoid unnecessary suffering. As many non-human animals are capable of suffering, we cannot mistreat them either. In Singer's words, 'If a being suffers there can be no moral justification for refusing to take that suffering into consideration.'[1] Thus, if someone accepts the principle of equality among humans, they cannot consistently reject that the same principle applies to non-human animals as well.

As a utilitarian philosopher, Singer maintains that the morality of an act depends on its consequences: an act is moral if it increases happiness, and immoral if it increases suffering. If an act causes more suffering than happiness, the act is morally wrong. As the interests of all individuals that have the capacity for suffering have to be given equal consideration, the suffering and the happiness of animals cannot be ignored. For example, if an act causes slight happiness for a few humans but results in severe suffering for many animals, the act is immoral.

---

[1] Singer, P. (1995): Animal Liberation. Pimlico, London, p. 8.

The importance of Singer's *Animal Liberation* lies not only in its ethical argument, but also in the fact that it exposes the cruelty of animal experiments and factory farms.

Most people think that researchers use and sacrifice animals in order to save human lives. However, this is usually not the case. In military experiments, animals are subjected to radiation, chemical warfare agents, or electric shocks, with the aim of inventing new methods to kill more humans in a more efficient way. Also, animals have to endure excruciating pain in experiments designed to test new cosmetics and household products. All of these experiments are unnecessary, for two reasons. First, we already have more than enough of cosmetics and household products, and introducing new ones will benefit no one, except the companies. Second, there are safe and cruelty-free ways of testing all these products, as demonstrated by a growing number of companies that have abandoned animal tests. As surprising as it may seem, even the scientific research using animals is all too often completely pointless. The results obtained are usually trivial or have been known for a long time. For a large number of studies it is obvious from the very first moment that they will never save a single human life or benefit humans in any way. These experiments only waste taxpayers' money and animals' lives.

Each year hundreds of millions of animals are used and killed in cruel and mostly needless experiments, many of which are performed without anaesthesia or pain relief. Animals are isolated, injected with viruses, starved, blinded, burned, drowned, mutilated, and dissected. In a just society, treating innocent beings in such merciless ways would be outlawed.

People usually do not make the connection between meat and the living beings who had to be killed to produce that meat. When people do make the connection, they tend to imagine a farm in the countryside, where cattle graze on the pasture, pigs root for acorns in the nearby grove, and chickens happily scratch around in the barnyard. Unfortunately, such idyllic farms are extremely rare. Most animals raised for food live in factory farms. A pig in a factory farm typically spends their entire life in boredom. They are sometimes confined to stalls where they cannot even turn around. Chickens are kept in crowded, windowless sheds and are slaughtered when they are 6 to 7 weeks old. Intensive egg production is even more cruel. First of all, males are not needed here, so the newly hatched male chicks are gassed, thrown into plastic bags where they will suffocate, or ground up alive. Laying hens undergo beak trimming without pain relief and are put into small cages – so small that the birds cannot even stretch their wings. The veal industry is no less abhorrent. Calves are separated from their mothers within hours of birth. In many countries they live alone in narrow stalls in which they cannot turn around. They are fed a liquid diet to keep their flesh pale and tender. Veal calves are slaughtered when they are a couple of months old. These are only a few of the cruelties associated with factory farms. Add all the other forms of mistreatment, including mutilations, the problems of transportation, and the horrors of slaughter, and you have a basic idea of what industrial animal farming is. Meat and other animal products are not essential for most humans. People may get some pleasure from eating meat, eggs, or dairy products, but this is outweighed by the intense suffering of the animals. Thus, factory farms are immoral.

When *Animal Liberation* was published in 1975, the phrase may have seemed a joke to many. Today Singer's argument is taken seriously even by those who do not agree with all of his conclusions.

In 1993 Paola Cavalieri and Peter Singer launched The Great Ape Project, the aim of which is to extend the rights to life, liberty, and protection from torture to chimpanzees, bonobos, gorillas, and orang-utans. 'If we regard human rights as something possessed by all human beings, no matter how limited their intellectual or emotional capacities may be, how can we deny similar rights to great apes?'[2] Singer asks. The recognition of our closest relatives as persons rather than mere 'things' would pave the way for an extension of the principle of equality to other animals beyond apes.

Utilitarianism judges the morality of an act by its consequences: an act is moral if it increases overall happiness, and immoral if it increases overall suffering. Utilitarianism is a major ethical direction, but there are different approaches, including deontology. Unlike utilitarianism, deontology does not focus on the net result of a given act. In other words, when deciding whether or not an act is moral, deontologists do not focus on the aggregate increase or decrease in happiness and suffering. Instead, proponents of deontology often rely on rights-based approaches. The difference between utilitarianism and deontology becomes evident if we imagine a situation in which sacrificing one individual would benefit a very large number of individuals. If the suffering caused by the act is outweighed by the increase in happiness, the act is not objectionable from a utilitarian perspective. Deontologists value the situation completely differently. If the individual has the right to life and protection from torture, this right cannot be overridden by the positive results the act would have on other individuals. Thus, not even one individual can be sacrificed for many others.

Ethics in favour of animals can be based on both utilitarian and deontological arguments. Tom Regan was a well-known proponent of the latter approach.

Tom Regan's 1983 book *The Case for Animal Rights* matches Singer's *Animal Liberation* both in its comprehensiveness and impact. To understand Regan's argument, we have to start from the inherent value of human beings. It is usually taken for granted that individual human beings have inherent value, that is, they are worthy of respect: humans should be treated as ends, never merely as means. Many think that humans have this inherent value because they are rational, are able to understand their duties, and can be held responsible for their actions. However, this view does not bear close inspection. Human infants, the senile, and the mentally disabled are not rational yet they have rights and are due respect. Or, to put it another way, they have inherent value. Regan argues that human infants, the senile, and the mentally disabled have inherent value because each one of them has a life that

---

[2] Singer, P. (2008): The Rights of Apes – and Humans. Project Syndicate, July the 15th, 2008. www.project-syndicate.org/commentary/the-rights-of-apes-and-humans?barrier=accesspaylog

matters to them. Their life can fare well or ill for them, which is independent of their utility for others.

Scientific evidence shows that several animals have very complex interior lives. They have beliefs and desires, perception, memory, a sense of the future, sentience, emotions, interests, a certain kind of autonomy, a psychophysical identity over time, and welfare. That is, each of them has a life that matters to them. Therefore, if humans have inherent value, we cannot deny that at least some animals have inherent value, too. This means that we have direct duties to them, or, to put it another way, they have rights. Individuals who have inherent value deserve to be respected, and we cannot harm them except under very special circumstances such as self-defence.

To sum it up, restricting moral considerability to humans is inconsistent and unjust. To treat animals merely as means is morally not acceptable. Consequently, exploiting animals for their meat, and killing or mistreating them for entertainment is wrong and should be abandoned. Also, research using animals is fundamentally wrong; the harms done to them in experiments cannot be offset, regardless of the potential benefits to humans. As we would never sacrifice a human in an experiment, so we cannot sacrifice an animal either.

The philosophical bases of Richard Ryder, Peter Singer, and Tom Regan are different, but the practical implications are overlapping. All three persons have had a great impact on how we treat animals. They have inspired millions to take action. Animal lovers all over the world realised that it is not enough to love their pets, but they have to do something for other animals as well. New animal advocacy organisations emerged, and the membership of the existing ones increased sharply. University courses started to discuss animal ethics, and the number of philosophical publications focusing on the issue grew rapidly. In some countries there have been important improvements in animal welfare legislation. Unfortunately, the fundamental characteristics of the systems that are based on the exploitation of animals have not changed. Factory farms and laboratories continue to view animals as raw materials, mere things without any moral significance. Deprived of everything that makes life worth living, animals have to endure the most barbaric treatments each day. They live in hell. Surely it was not without reason that Yiddish writer and Nobel laureate Isaac Bashevis Singer compared the situation to the death camps of the Nazis, writing about 'an eternal Treblinka'[3] for the animals. But it does not have to be eternal. Certainly not overnight, but we, as conscientious citizens can challenge and change the way animals are treated.

---

[3] Singer, I. B. (2004): Collected Stories: Gimpel the Fool to the Letter Writer. The Library of America, p. 750.

Worth Reading

Cavalieri, P., & Singer, P. (1994). *The great ape project: Equality beyond humanity*. New York: St. Martin's Press.

Mason, J., & Singer, P. (1990). *Animal factories*. New York: Three Rivers Press.

Regan, T. (2004a). *The case for animal rights*. Berkeley: University of California Press.

Regan, T. (2004b). *Empty cages: Facing the challenge of animal rights*. Lanham: Rowman & Littlefield Publishers.

Regan, T. (2006). *Defending animal rights*. Urbana: University of Illinois Press.

Ryder, R. D. (1983). *Victims of science: The use of animals in research*. London: National Anti-Vivisection Society.

Ryder, R. D. (2001). *Painism: A modern morality*. London: Centaur Press.

Ryder, R. D. (2011). *Speciesism, painism and happiness: A morality for the twenty-first century*. Exeter: Societas.

Singer, P. (2009). *Animal liberation*. New York: Harper.

Singer, P., & Mason, J. (2006). *The ethics of what we eat: Why our food choices matter*. Emmaus: Rodale.

Worth Browsing

petersinger.info
www.greatapeproject.uk

Worth Watching

Speciesism: The Movie (2013)
The Animals Film (1981)
Tools for Research (1983)
From Mice to Men? (2007)
We Are All Noah (1986)
Live and Let Live (2013)
Empathy (2017)

How Henry Spira Put Animal Liberation into Practice

In the 1970s animal ethics entered the sphere of academic dialogue. Also, it motivated an increasing number of citizens to start translating philosophy into action. Henry Spira was one of the greatest among these new activists.

Henry Spira was born in Antwerp, Belgium, in 1927. In 1938 the family migrated to Panama but two years later they moved on to the USA, where they settled in New York. In 1945 Spira became a merchant seaman. In 1953 he was drafted into the army, where he served two years. After his discharge he began working in a car factory. Meanwhile he was studying at Brooklyn College, graduating in 1958.

As a young boy Spira became interested in socialist political theories. He was involved in the activity of trade unions, for which he was blacklisted as a security risk. Spira also participated in the protests against the segregation laws that treated African-Americans as second-class citizens. In 1966 Spira started to teach in a high school. His students were kids from ghettos. Spira enjoyed the job and used innovative methods to motivate the children.

Henry Spira had an extraordinary ability to identify with the vulnerable. But animal advocacy was beyond his interest until 1973. In that year, one of his friends travelled to Europe and asked Spira to look after a cat named Savage. Spira was not particularly happy about the idea but did not want to refuse. It didn't take long for the cat to enchant Spira. The new friendship had a profound impact on Spira's thinking. 'I soon began to wonder about the appropriateness of cuddling one animal while sticking a knife and fork into others,'[1] Spira wrote.

It was about this time that Spira came across a newspaper article that tried to ridicule the 'animal liberation' standpoint of philosopher Peter Singer. Spira read the original article of Peter Singer and thought that Singer's argument did make sense. 'Singer described a universe of more than 4 billion animals being killed each year in the USA alone. Their suffering is intense, widespread, expanding, systematic and socially sanctioned. And the victims are unable to organize in defence of their own

---

[1] Spira, H. (1985): Fighting to Win. In Singer, P. (ed.) In Defense of Animals. Basil Blackwell, New York, pp. 194–208.

© Springer Nature Switzerland AG 2019
L. Erdős, *Green Heroes*, https://doi.org/10.1007/978-3-030-31806-2_11

interests. I felt that animal liberation was the logical extension of what my life was all about – identifying with the powerless and the vulnerable, the victims, dominated and oppressed,'[2] Spira recalled his impressions when reading the essay. In 1974 he enrolled in Singer's evening course about animal liberation at New York University.

Once Spira understood that the exploitation of animals is morally wrong he wanted to do something about it. First, he became a vegetarian. Second, with a handful of other animal advocates, he established Animal Rights International (ARI) and started to think about campaigns to help animals. Spira felt that the ethic of animal liberation calls for radical changes in each system that is based on the use and abuse of animals, including science, agriculture, and the cosmetics industry. However, based on his experience in the civil rights movement, he knew that he had to focus on a single problem, setting a goal that is achievable within a realistic time frame.

ARI's first target was the American Museum of Natural History's experiment on cats. Based in New York, the Museum was mutilating cats and observing how this affected their sexual behaviour. Spira tried to contact the Museum but his requests for a meeting were ignored. Spira published articles and appeared in interviews, criticising the pointless experiment. ARI organised protests: animal lovers demonstrated in front of the Museum each weekend for one and a half years. Also, citizens were encouraged to phone the Museum and to write letters demanding an end to the cruelty. The campaign attracted the attention of politicians and scientists. Congressman Ed Koch visited the laboratory to find out about the expected benefits of the study but the museum officials couldn't mention a single point. Even the scientific community denounced the experiment as useless. As a result of the campaign, the funding for the experiment was cut in 1977. The torture of dozens of cats was ended, but the victory also had a psychological importance, as it showed animal advocates that, with concentrated efforts, they are able to accomplish significant results.

In the late 1970s Spira challenged the Pound Seizure Law of New York State, which made it possible for research institutions to acquire dogs and cats from shelters for use in experiments. The cheap and seemingly unlimited source of lab animals encouraged the researchers to use them as if they were disposable laboratory equipment. Spira teamed up with other animal advocates and started to lobby state legislators. Thanks to the efforts, the law was repealed in 1979.

Spira's next campaign was against the Draize test, which involves placing various chemicals into the eyes of animals (mostly rabbits) in order to test the damages the chemical does to the eye tissues. Imagine how terrible it is to have some irritating substance in your eyes for just a couple of minutes. Test animals have to endure this for days, until their eyes are seriously damaged. The Draize test was widely used in the cosmetics industry. Spira needed a concrete target, which turned out to

---

[2] Spira, H. (1985): Fighting to Win. In Singer, P. (ed.) In Defense of Animals. Basil Blackwell, New York, pp. 194–208.

be Revlon, one of the most influential cosmetics companies. Spira contacted Revlon and asked them to fund research to develop alternatives to animal tests. The representatives of Revlon were ready to negotiate but were not willing to accept Spira's proposal. Once again, Spira turned to the public. ARI published full-page ads in several newspapers, including *The New York Times*, asking 'How many rabbits does Revlon blind for beauty's sake?' When letters began pouring in to Revlon by the thousands, the management changed their mind. Revlon donated a substantial sum to explore cruelty-free alternatives to animal testing. Not to be outdone, other major companies followed the example, boosting the research into alternatives to animals in product safety testing. Today there is a wide selection of cruelty-free products and many companies do not use animal tests any more. However, several companies have not abandoned testing on animals, or have recently resumed animal tests, either because it is cheaper than alternatives or because they entered the Chinese market, where tests on animals are compulsory. Unfortunately, the Draize test is still used in laboratories, although the method has been refined and the number of animals used has been reduced.

In the early 1980s Spira focused his attention on the LD50 test. An abbreviation for lethal dose 50%, the test is used to examine the toxicity of cosmetics and household products by finding the amount that kills half of the study animals. The material (anything from shampoo to detergent to pesticide) is forced down the throat of rats, mice, cats, dogs, or any other available animals. Alternatively, the substance can be applied on the skin or injected to the animal. The test results in severe pain, bleeding, diarrhoea, and usually an agonising death. When Spira started the campaign it soon turned out that a great number of toxicologists considered the LD50 test to be misleading and, consequently, needless. As a result of Spira's efforts, there has been a sharp reduction in the number of animals used in LD50 tests, although it has not yet been totally abandoned.

While experiments using animals are causing the suffering and death of hundreds of millions of animals worldwide each year, the number of animals slaughtered for food is two or three orders of magnitude higher: up to 150 billion animals are killed annually for meat. Moreover, most of them are kept in factory farms and thus have no chance of a decent life. That is why Spira turned his attention to industrialised farming. Spira connected the animal advocacy issues with the factory farms' terrible workers' safety records, the unhealthy end products, and the serious environmental problems. Spira was convinced that the best solution would be if humans were not eating meat. ARI was working intensively to reduce meat consumption. At the same time they tried to achieve improvements in the animals' living conditions. A related campaign targeted the face-branding of the cattle imported from Mexico. Face-branding was used to mark cattle for regulatory purposes. Animal Rights International contacted the Department of Agriculture to change the regulations but the attempts were to no avail. Spira's team decided to raise public awareness about the issue. When they published ads about how cruel the practice was, concerned citizens flooded the authorities with hundreds of phone calls and thousands of letters. As a result, face-branding was banned. In a subsequent advertisement, ARI thanked the Department of Agriculture for the change.

In 1995 Spira was diagnosed with cancer. He remained positive and, despite his declining health, continued to work for animals right to the last moment. Henry Spira passed away in 1998.

Spira kept bureaucracy to the minimum. His apartment served as the 'office' of Animal Rights International, and he did not have to maintain a large staff. Though Spira did have a few helpers, ARI was largely a one-man organisation. This made ARI extremely cost-effective, but it also had a drawback: after Spira's death the organisation's activity declined and in 2010 it was discontinued.

The success of Spira's campaigns was based on multiple factors. First and foremost, he had a gradualist approach with attainable goals. He carefully prepared for each campaign, meticulously collecting all available information on the issue. As a result, he was able to back his claims with accurate data and could offer viable solutions to reduce animal suffering. He was always willing to negotiate, but, if necessary, he was also able to generate popular support for the cause. He was perseverant and never gave up fighting. But most importantly, Spira had a solid moral background for everything he did: 'The other side has the power, but we have justice on our side.'[3]

The Strategy of Small Steps and The All-or-Nothing Approach

The strategy of Henry Spira received some criticism because his willingness to negotiate was occasionally interpreted as a betrayal of the ideology of the animal advocacy movement. Nevertheless, Spira's strategy seems to have a lasting effect and continues to influence the activity of many animal advocacy groups.

Spira's campaigns had relatively modest short-term objectives, but he never lost sight of his long-term vision. He envisioned a world in which animals are not enslaved, not exploited, not killed, and not eaten. However, he knew that great changes usually do not occur overnight. He valued small victories because these were able to save the lives of many animals. Also, Spira was convinced that small results bring animal liberation closer one step at a time.

Realising that a few small victories are better than a huge defeat, Spira set limited goals and focused on specific and easily identifiable targets. For example, he opposed all types of vivisection, but he 'only' campaigned against certain types at specific locations at a time.

Today many activists think that combining long-term vision and realistic short-term objectives can mean an immediate help for a large number of animals without compromising the basic principles of animal advocacy.

---

[3] Spira, H. (1985): Fighting to Win. In Singer, P. (ed.) In Defense of Animals. Basil Blackwell, New York, pp. 194–208.

Henry Spira saw the animal advocacy movement as a continuation of the women's and civil rights movements. Just as we cannot oppress or enslave those who happen to belong to another sex or race, so we cannot oppress or enslave those who happen to belong to another species. According to Spira, the purpose of our life is to help all, humans and non-humans, whose rights are withheld. As Spira said, 'What greater motivation can there be in a person's life than doing anything one possibly can to reduce pain and suffering?'[4]

Worth Reading

Singer, P. (1998). *Ethics into action: Henry Spira and the animal rights movement*. Lanham: Rowman & Littlefield Publishers.

Worth Watching

Henry: One Man's Way (1997)

---

[4] Henry Spira in the 1997 film Henry: One Man's Way.

Ingrid Newkirk, Alex Pacheco, and PETA

'Animals are not ours to experiment on, eat, wear, use for entertainment, or abuse in any other way.'[1] This is the motto of People for the Ethical Treatment of Animals (PETA), the world's largest and best-known animal advocacy organisation, which was started by two devoted activists: Ingrid Newkirk and Alex Pacheco.

Ingrid Newkirk was born in Surrey, England, in 1949. When she was 7 years old, the family moved to New Delhi, India, where her father worked as a navigational engineer, while her mother volunteered for Mother Teresa. Ingrid helped her mother with caring for people in need. Also, it was during this time that she started her career as an animal activist, feeding stray animals on the streets of New Delhi.

In 1967 Ingrid's family settled in the US. She decided to become a stockbroker, but an unexpected life-changing event turned her into an animal advocate. When one of her neighbours moved away and abandoned almost a dozen cats, Ingrid collected the animals and took them to the nearest shelter. At the shelter she handed over the cats and, after a couple of minutes, asked if she could see them again. She was shocked to learn that the cats had already been killed.

Ingrid found the conditions at the shelter so bad that she wanted to do something about it. She immediately abandoned her career in brokerage and took up a job at the shelter. She had to witness acts of senseless cruelty each day. She would report the incidents to the management, but no one cared. For a couple of years Ingrid worked as an animal-protection officer. During this time she realised that cruelty to pets is basically not different from the cruelty animals raised for food have to endure. Buying meat means supporting animal cruelty. Ingrid Newkirk became a vegetarian. In 1980 she met Alex Pacheco, a part-time volunteer at an animal shelter. The encounter had a profound effect on the history of animal advocacy.

Originally, Alex Pacheco wanted to become a Catholic priest. In 1978 he visited a slaughterhouse where one of his friends had a summer job. The event was a turn-

---

[1] www.peta.org

© Springer Nature Switzerland AG 2019

L. Erdős, *Green Heroes*, https://doi.org/10.1007/978-3-030-31806-2_12

ing point is his life. After Pacheco saw 'the violent deaths of terrified dairy cows, pigs, and chickens,'[2] he was determined to join the animal advocacy movement.

Newkirk and Pacheco thought that the most efficient way of working for animals would be to start a radical organisation. In 1980 they set up People for the Ethical Treatment of Animals. The newly formed group turned its attention to vivisection and started an operation that eventually became one of the best-known actions of PETA.

Pacheco applied for a job at the Institute for Behavioral Research in Silver Spring, Maryland. The Institute was conducting experiments that included cutting selected nerves of the monkeys, then using various treatments such as electric shocks, and observing the monkeys' behaviour. At the time there was no vacancy, but Pacheco was offered a volunteer position. What Pacheco saw in the lab would perfectly fit a horror movie. There was dirt, faeces, blood and urine everywhere. Rats and cockroaches were roaming the facility. Starving, neurotic monkeys were confined to small metal cages in a windowless room; they received no bedding, no environmental enrichment, and no veterinary care. 'It was a nightmare,'[3] Pacheco said.

It was difficult for Pacheco to return to the lab every day, but he wanted to document the conditions. He worked like a secret agent. As time went on, the laboratory personnel started to trust Pacheco. He was allowed to work alone, received his own set of keys, and could enter the lab at night and on weekends. He discreetly helped the animals with some additional food, and he took photographs. In addition, he organised a secret tour in the facility for experts who could testify to the negligence and cruelty in the lab.

With the evidence and the expert's affidavits, Newkirk and Pacheco turned to the authorities. The police raided the institute, which was quickly picked up by the media. A long legal battle unfolded for the custody of the monkeys. Unfortunately, the animal advocates' efforts to send them to a sanctuary failed. Some of the monkeys were sent to a zoo, the less lucky ones had to endure further experiments, after which they were killed. The head of the horror lab was convicted but later acquitted, eventually getting off unpunished.

Even so, PETA's campaign was not a complete failure. Millions of citizens became aware of what was going on behind the closed doors of laboratories. Also, animal abusers had to reckon with further infiltrations. Indeed, infiltration has remained one of the main components of PETA's tactical repertoire ever since, and is regularly used to document violence in laboratories, slaughterhouses, factory

---

[2] Pacheco, A.; Francione, A. (1985): The Silver Spring Monkeys. In Singer, P. (ed.) In Defense of Animals. Basil Blackwell, New York, pp. 135–147.

[3] Changing Times, Changing Minds: A PETA Family Album. https://www.peta.org/videos/changing-minds-changing-times-a-peta-family-album/

farms, roadside zoos, circuses, and fur farms. Some of these undercover operations have been quite successful and managed to save a large number of animals.

In 1984 the Animal Liberation Front sent PETA a videotape showing extremely cruel experiments that were being carried out in a laboratory at the University of Pennsylvania. Not only did the footage reveal how brutally the monkeys were treated but it also showed how the researchers made fun of the situation, laughing at the animals' suffering. PETA and other animal advocacy groups started to organise demonstrations. On a summer's day in 1985, Ingrid Newkirk and Alex Pacheco, accompanied by 99 other activists, walked into the funding agency's office, demanding that the grants be withdrawn from the research. As the sit-in made the headlines, supporters and journalists gathered around the building. On the fourth day of the protest, the grant was suspended and an investigation was started. Due to serious violations of animal welfare regulations, the laboratory was closed.

PETA has achieved notable improvements in animal welfare legislation, including a ban on shooting live animals in military weapons tests. Also, the practice of using live animals in car crash tests ended due to the efforts of PETA. The organisation has managed to persuade major cosmetics companies to stop testing on animals, fashion designers not to use fur, restaurant chains to improve animal welfare requirements, and retailers to stop selling glue traps.

Each day PETA field workers are carrying out significant activity to help neglected and abused pets and farm animals. They visit poor neighbourhoods, talk with the owners of the animals, and provide them with doghouses and low-cost spay and neutering service.

PETA publishes books, distributes leaflets, and releases short videos in order to encourage people to adopt a cruelty-free lifestyle. One of their best-known campaigns informs people about how farm animals are raised and killed. Animal advocates believe that if people knew what is happening in factory farms and slaughterhouses, most of them would become vegetarian or vegan. PETA emphasises that plant-based diet is not a set of inconvenient restrictions. On the contrary, it opens up a whole new world of previously unexplored fruits, vegetables, and seasonings. Vegetarianism and veganism are not only compassionate, but also exciting ways of living.

Some Famous Vegetarians

People can adopt a vegetarian diet for numerous reasons. Some want to improve their health, others are concerned about the terrible environmental impacts of producing meat. Here I mention some notable figures whose primary reason for becoming vegetarian was their compassion for animals.

Renowned Russian author Leo Tolstoy was strongly convinced that eating meat is morally bad, as it entails the killing of animals. According to Tolstoy, 'A man can live and be healthy without killing animals for food; therefore, if he eats meat, he participates in taking animal life merely for the sake of his appetite. And to act so is immoral.'[4]

Nobel laureate playwright George Bernard Shaw said, 'Animals are my friends, and I don't eat my friends.'[5]

For Mahatma Gandhi, the leader of the Indian independence movement, the principle of non-violence applied to humans and non-humans alike, thus he was a strict vegetarian. He regarded the conditions under which dairy cows are reared as morally unacceptable, therefore, he drank the milk of his own goat.

Primatologist and activist Jane Goodall adopted vegetarianism after she had read Peter Singer's book *Animal Liberation*.

Former Beatles bass guitarist and singer Paul McCartney and his wife Linda decided to give up eating meat in the mid-1970s. One day they were eating lamb chops when they looked out of the window and saw sheep and lambs. They suddenly made the connection. 'It was like, the penny dropped. The light bulb lit up,'[6] McCartney said.

Queen guitarist Brian May is a devoted vegetarian. In an interview he said, 'To me every creature on this planet has an equal right to have a good life.'[7]

Oscar-winning actress Natalie Portman has been a vegetarian since she was 9 years old, and adopted veganism in 2011. She produced and narrated the 2017 documentary *Eating Animals*.

The simplest and most effective way of helping animals is not to eat them. A vegetarian saves dozens or even hundreds of animals each year, the accurate number depending on many factors such as location and exact diet type. And it's not just about saving the animals' lives; it's also about not paying for acts of cruelty that are the norm in many farms, especially factory farms.

---

[4] Quoted in Lashley, D. (2018): Clean Food for Clean People. Lulu Publishing Services, p. 91.

[5] Quoted in Cox, P. (2002): You Don't Need Meat. St. martin's Press, New York, p. 113.

[6] Interview with Paul McCartney, The Guardian, July the 18th, 2010. https://www.theguardian.com/music/2010/jul/18/paul-mccartney-vegetarianism

[7] Brian May on KLOS Jonesy's Jukebox, August the 24th, 2017. https://brianmay.com/brian/brian-news/briannewsaug17.html

On-line petitions are among the most effective forms of PETA's work. It takes only a few seconds to sign a petition, and if there are thousands or tens of thousands of citizens who join a campaign, together they are usually able to evoke changes in legislation or corporate culture.

PETA has a very large number of celebrity supporters. Pamela Anderson, Alec Baldwin, Natalie Portman, Alicia Silverstone, and Paul McCartney are just a few of the most devoted animal lovers who regularly appear in the organisation's campaigns.

These days it is difficult to attract people's attention. But photos showing nude celebrities are able to reach a huge audience, and PETA does not miss this opportunity to convey animal advocacy messages. The anti-fur project with the slogan 'I'd rather go naked than wear fur' has been joined by many a famous singer, model, actor, and sportsperson, all of whom were ready to strip down in order to help animals.

PETA is sometimes criticised for publishing materials that are disturbing. No one really likes to watch how animals are being tortured and PETA activists certainly do not enjoy making and distributing these videos. However, we have to face reality, and it is not PETA's responsibility that atrocities and violations of animal welfare regulations are so widespread. Rather than deploring PETA, critics should condemn those who routinely torture animals.

PETA is usually reproached for operating open admission shelters, which means that the organisation does not implement a no-kill policy. No-kill shelters refuse to take in new animals when the facility is full, so they do not have to kill healthy or treatable animals. However, according to PETA, this generates more problems than it solves, as the animals who were not admitted will either die in agony on the street, or produce offspring, thereby exacerbating overpopulation. It would be hard to estimate which type is more humane. What is sure is that the long-term solution of the problem would be to avoid overpopulation. To achieve this, PETA popularises adoption from shelters instead of buying from breeders, and runs spaying and neutering programmes.

Some of those who oppose the activity of PETA claim that Ingrid Newkirk is a misanthrope. However, this is not true. Newkirk has witnessed the darkest side of human behaviour, but she also knows how compassionate and helpful people can be. And she does not place animals above humans; all she says is that the suffering of animals is morally no less important than the suffering of humans, which is a defensible position.

To some people, PETA's vision of abolishing the exploitation of animals seems abhorrent. The view that animals are morally not considerable is so deeply ingrained in our societies that it is difficult to challenge the use and abuse of non-human animals. However, thanks to the activity of PETA and other animal advocacy groups, more and more people are willing to accept that, after all, 'Animals are not ours to experiment on, eat, wear, use for entertainment, or abuse in any other way.'

Worth Reading

Newkirk, I. (1999). *You can save the animals: 251 simple ways to stop thoughtless cruelty.* Rocklin: Prima.

Newkirk, I. (Ed.). (2002). *The PETA celebrity cookbook: Delicious vegetarian recipes from your favorite stars.* New York: Lantern Books.

Newkirk, I. (2005). *Making kind choices: Everyday ways to enhance your life through earth- and animal-friendly living.* New York: St. Martin's Griffin.

Newkirk, I. (2006). *50 awesome ways kids can help animals: Fun and easy ways to be a kind kid.* Boston: Warner Books.

Newkirk, I. (2009). *The PETA practical guide to animal rights.* New York: St. Martin's Griffin.

Newkirk, I. (2012). *Free the animals! The amazing true story of the animal liberation front in North America.* New York: Lantern Books.

PETA, & Newkirk, I. (1993). *The compassionate cook: Or, please don't eat the animals!* New York: Warner Books.

Worth Browsing

www.peta.org
www.peta.org/action/

Worth Watching

The 8 Secrets of How PETA Works (2014)
Nonviolence Includes Animals (2014)
I Am an Animal: The Story of Ingrid Newkirk and PETA (2007)
The Age of Beasts (2017)
Changing Minds, Changing Times: A PETA Family Album (1991)

Talking Apes – Ambassadors of the Animal Kingdom in the Human World

Throughout history humans have been intrigued by the idea of talking with animals. But for millennia this was deemed impossible by scientists and philosophers, most of whom regarded animals as mindless beasts. With Darwinian evolution it became evident that there is no gulf between humans and other animals. Consequently, it seemed logical to assume that some animals may also possess linguistic skills. In the twentieth century pioneering studies aimed to find out whether non-human apes can be taught to talk. Chimpanzees, our closest relatives, were tested first.

Initial observations were quite disappointing. In the 1940s, after 6 years of training, Viki, a female chimpanzee could speak only four words. However, two brilliant scientists, Beatrix and Allen Gardner were convinced that the failure was not due to the low chimpanzee intelligence. Chimps are anatomically not well equipped to produce intelligible speech. In addition, they are more interested in gestures than in voices. In other words, chimpanzees are evolutionarily prepared for non-verbal rather than verbal communication. Therefore, the Gardners decided to teach American Sign Language (ASL) to a chimpanzee.

Washoe was a female chimp who had been captured in Africa in the mid-1960s, a time when it was legal to import wild-caught chimpanzees into the US. She was intended for aeromedical research, but was fortunate enough to be selected by the Gardners for their linguistic experiment. The project started in 1966. Washoe was raised like a human child, in a warm and loving family environment. It quickly became evident that Washoe was able to learn the signs of ASL. It was also clear that she could generalise one sign to several similar objects, for example, she understood that the sign for 'drink' means not just one specific type of drink but many types such as water or soda pop. Before long Washoe started to combine signs into phrases, grasped grammatical rules, and began using simple sentences. Washoe was the first non-human who acquired a human language.

Washoe was a charming little girl. She liked playing with dolls and leafing through books and magazines, loved hide-and-seek, and enjoyed bedtime stories – of course, in sign language. And, like most children, she was quite rambunctious at

© Springer Nature Switzerland AG 2019

L. Erdős, *Green Heroes*, https://doi.org/10.1007/978-3-030-31806-2_13

times. Washoe liked car rides, but for some mysterious reason she became furious whenever she spotted a motorcycle policeman.

From 1970 Roger Fouts became Washoe's main tutor. He continued the observations and involved additional chimps in the study. The chimpanzees started spontaneously communicating in sign language among themselves. What was probably even more exciting form a scientific perspective, it also turned out that ASL was passed from one chimp generation to the next.

The Gardners and Roger Fouts, with the help of their chimpanzee friends, demonstrated the continuity between man and non-human animals and shed light on the evolution of language. Also, they have shown that research involving animals can be done humanely while still maintaining a high scientific standard. However, despite the groundbreaking results, Fouts had to deal with a personal crisis. After visiting numerous primate labs across the US, he had to face the fact that chimpanzees were being kept under extremely cruel conditions, locked up in tiny cells, with no natural light, no toys, no blankets, and no companions. They were subjected to painful or even deadly experiments. Some were injected with viruses and hermetically sealed in sterile isolation chambers. Fouts' studies were neither painful nor lethal, and his chimpanzees had happy lives with social contacts and an enriched environment. They received the best possible care. Still, Fouts felt he had been 'part of a system that was breeding more chimpanzees for more suffering.'[1] He decided to quit that system.

Fouts started a struggle with the long-term goal of gradually phasing out all research involving captive animals. Fouts meant all such research, including his own studies. As the first step in this long battle he began to fight for improved conditions for lab animals. He proposed that larger laboratory cages be used, and the animals be provided with social contacts, toys, and mental activities. Fouts also started to advocate seeking alternatives to animal experiments whenever possible, plus terminating experiments that are totally meaningless from a scientific perspective.

Washoe passed away in 2007. Roger Fouts, inspired by his work with Washoe, has done a huge work for chimpanzees and other animals. He has published articles challenging the cruel treatment of laboratory animals, started educational programmes for students, and negotiated with scientists, politicians, and zookeepers. He has long been an advocate of building large and humane refuges for chimpanzees who are not needed any more in research. 'Over the past 40 years we have spun chimpanzees in centrifuges and shot them into space. We have smashed their skulls with steel pistons and used them as crash test dummies. We have deprived them of all maternal contact and driven them to psychosis. We have used them to test lethal pesticides and cancer-causing industrial solvents. We have injected them with massive doses of polio, hepatitis, yellow fever, malaria, and HIV,'[2] Fouts wrote. And he

---

[1] Fouts, R.; Mills, S. T. (2003): Next of Kin: My Conversations with Chimpanzees. Harper, New York, p. 204.

[2] Fouts, R.; Mills, S. T. (2003): Next of Kin: My Conversations with Chimpanzees. Harper, New York, p. 364.

insists that those who managed to survive deserve a decent life for their remaining years.

Washoe taught us that assuming the existence of a chasm between humans and the rest of the animal world had been a fatal error. As Fouts wrote, 'the chimpanzee mind and the human mind are fundamentally alike.'[3]

Shortly after Washoe's project had started, a similar study was launched by Ann and David Premack. The research involved a chimpanzee named Sarah, who was taught an artificial language, in which words were represented by differently shaped and coloured pieces of plastic. Sarah learned a considerable vocabulary of nouns, verbs, adjectives, and pronouns. She grasped grammatical rules and was able to understand compound sentences. The study ended when Sarah was about 10 years old. She passed away in 2019.

Bonobos (also called pygmy chimpanzees) have also demonstrated their ability to learn a human language. Probably the most famous of them has been Kanzi, a male bonobo. Kanzi's incredible story began when he was an infant. Psychologist Sue Savage-Rumbaugh tried to teach an artificial language to Matata, Kanzi's adoptive mother. The efforts seemed quite fruitless as Matata did not comprehend the idea. However, to the researchers' astonishment, Kanzi, who usually played in the proximity during the lessons, learned the words (represented by symbols) effortlessly. Once he became the primary subject of the study, Kanzi proved to be a diligent pupil. He has learned hundreds of symbols, and also understands thousands of spoken English words. Moreover, he has picked up ASL signs from watching videos of Koko, a gorilla who was taught to communicate using sign language.

Kanzi is able to cook his dinner, wash the dishes, and light a campfire. He likes to skype and watch DVDs, and loves puzzles and mental challenges. Videos of Kanzi working in the kitchen, doing a language test in the lab, or starting a fire show how surprisingly human he is.

Based on her findings, Sue Savage-Rumbaugh thinks that the sharp line that was once assumed to separate humans and non-human animals simply does not exist. And this has obvious moral implications on how we treat apes. Savage-Rumbaugh maintains that all harmful biomedical research on apes must be banned.

Nim Chimpsky has been one of the most famous signing chimpanzees. Interestingly, after working with Nim, psychologist Herbert Terrace concluded that Nim used signs but he only imitated humans, thus his performance did not qualify as language. Some experts think that Terrace's findings were based on his erroneous study design, dated methods, and incorrect data evaluation. A subsequent study with Nim suggested that, contrary to Terrace's claims, Nim did use a simple form of language. In 1982 Nim was sold to a laboratory. Animal lovers across the US were shocked to learn that this famous signing chimp was transferred to a biomedical lab, where he was kept in a solitary cage. In the next year, Cleveland Amory, the famous writer and animal advocate purchased Nim from the laboratory and brought him to

---

[3] Fouts, R.; Mills, S. T. (2003): Next of Kin: My Conversations with Chimpanzees. Harper, New York, p. 344.

his Black Beauty Ranch, an animal sanctuary Amory had established a few years earlier. Nim spent the next 17 years of his life on the farm, where he passed away in 2000. His biography, written by Elizabeth Hess and published in 2008, became a bestseller and an important book examining our relation to our closest living relatives.

Ape language research flourished in the 1970s. Besides chimpanzees and bonobos, gorillas and orang-utans quickly became the subjects of these studies. Francine Patterson started to teach sign language to Koko, a female gorilla in 1972. According to Patterson, Koko mastered a vocabulary of over 1000 signs and was able to understand some 2000 spoken English words. She understood grammatical rules and invented new combinations (i.e., compound words) by creatively using existing signs. For example, Koko was not taught the sign for 'ring.' Therefore, Koko combined the signs 'finger' and 'bracelet' into a compound word referring to ring. Koko was able to use her linguistic skills to communicate about complex thoughts, feelings, and emotions.

Koko loved to talk with her caretakers. Her other hobby was watching movies; *Free Willy* was her favourite. Koko's pastime activities included eating. She especially enjoyed her birthday parties, when she was served her favourite fruits and juices. Koko liked to receive guests at her home. Her visitors, including notable celebrities such as Robin Williams and Leonardo DiCaprio, were always amazed by her friendliness.

Despite her tremendous strength, Koko was a gentle and caring person. In 1983 Koko asked for a cat for her birthday. When she received a toy cat, she was terribly disappointed. Obviously, she had meant a real cat. So Koko was given a kitten who had been abandoned by her mother. Koko named the kitten All Ball. The huge gorilla played with the kitten in such a gentle way that the tiny All Ball had nothing to fear from Koko. The unusual friendship made news worldwide but ended untimely when the cat was run over by a car. Patterson's award-winning children's book *Koko's Kitten* is dedicated to the memory of All Ball.

Koko died in 2018, at the age of 46. She was arguably the best-known gorilla in the world: she appeared on the cover of *National Geographic*, featured in several documentary films, and was the topic of countless conference speeches, articles, and books. Thanks to her intelligence and engaging personality, Koko was a real star with millions of fans all over the world. Koko's memory and influence lives on. Francine Patterson and The Gorilla Foundation, co-founded by Patterson in 1976, continue to fight for the welfare of captive gorillas as well as the conservation of wild-living populations of all ape species.

In the second half of the 1970s Lyn Miles started a study at the University of Tennessee at Chattanooga to examine the linguistic ability of orang-utans. She taught sign language to a male orang-utan named Chantek. The young orang attended sign language lessons each day and before long he used some 200 signs. He created some new combinations. For instance, he used the signs 'eye-drink' to refer to the caretakers' contact lens solution, and the combination 'tomato-toothpaste' for ketchup. It was also clear that Chantek had the ability to refer to objects that were not present. Surprising as it may seem, Chantek could even tell lies. In addition to the use of sign language, Chantek was also able to comprehend spoken English. He was so respected and beloved at the university campus that his photo was included in the university's yearbooks.

Chantek learned to use human tools such as screwdrivers, and enjoyed activities that represented mental challenges for him. He understood the concept of money. The results of Miles' study indicate that Chantek might have been able, at a basic level, to distinguish between morally good and wrong acts.

When Chantek grew up he became so large and powerful that it was difficult to control him. He was sent back to the Yerkes Primate Center, where he had been born. Being kept in a very small cage, he became depressed. When he was visited by his previous caretakers, he would sign that he wanted to be freed and go home. In 1997 Chantek joined a group of orangs in a large and enriched enclosure at Zoo Atlanta, where he died in 2017.

There are still some who dispute the linguistic capacity of great apes. But no one can doubt that great apes are highly intelligent beings with complex mental and emotional lives. And these facts have some inescapable moral implications. Clearly we cannot continue to use our closest living relatives in harmful experiments. We have a moral obligation to ensure a decent life for those who are already in captivity, we must stop breeding them, and every effort has to be made to protect the remaining wild populations in Africa and Southeast Asia.

Worth Reading

Fouts, R., & Mills, S. T. (2003). *Next of kin: My conversations with chimpanzees*. New York: Harper.

Hess, E. (2008). *Nim Chimpsky: The chimp who would be human*. New York: Bantam Books.

Patterson, F., & Linden, E. (1981). *The education of Koko*. New York: Holt, Rinehart & Winston.

Premack, D., & Premack, A. J. (1983). *The mind of an ape*. New York: Norton.

Savage-Rumbaugh, S., & Lewin, R. (1994). *Kanzi: The ape at the brink of the human mind*. New York: Wiley.

Worth Browsing

www.friendsofwashoe.org
www.releasechimps.org
www.koko.org
www.greatapeproject.uk

Worth Watching

Kanzi: An Ape of Genius (1993)
Project Nim (2011)
Koko, the Gorilla who Talks (2016)
A Conversation with Koko (1999)
My Wild Affair – The Ape Who Went to College (2013)

Parrots, Dolphins, Seals – And What They Have Taught Us

Unlike the great apes, parrots can imitate a wide variety of sounds, including human speech. However, they are usually only parroting: they mimic our words but do not understand what these words mean. Could they be taught to talk meaningfully? For a long time this seemed totally impossible. Parrots have bird brains that are large relative to their body size but very small compared to the brains of great apes. In addition, the avian brain's organisation is very different from that of primates. In the 1970s advanced cognition seemed far beyond the capacity of bird brains. But Irene Pepperberg thought otherwise.

Irene Pepperberg graduated as a chemist and earned her PhD in chemical physics at Harvard University in 1976. Her career changed course in 1977, when she bought a male African grey parrot and started a study examining avian cognition. She named the parrot Alex, which is an acronym for Avian Learning Experiment. The project was met with scorn and resistance. Irene Pepperberg had to make great personal sacrifices in order to continue the work: she faced financial difficulties, unemployment, and ridicule. But she persevered and eventually her efforts paid off.

Pepperberg applied the model/rival approach in the training because she felt this would fit with the sociability and natural curiosity of the parrots. The technique applies two human trainers who talk to each other, and the bird has to compete for the trainers' attention and for small rewards. The method proved very fruitful indeed. Alex learned over a hundred English words and was able to answer questions and make requests. He could communicate about the colour, shape, material, and size of several objects. He understood the concept of sameness and difference as well as the concept of 'none.' Using his existing vocabulary he was able to form new combinations. Alex had some mathematical knowledge as well: he could count to 6 and add up small numbers. He also participated in the training of other grey parrots. All in all Alex had the intelligence of a five- or six-year-old child.

Alex had a rather bossy personality: he had to be greeted first in the morning, otherwise he refused to work with Pepperberg. Sometimes he would not co-operate with the trainers if he found the task boring. And he usually became angry when he did not get what he asked for. As Pepperberg noted, Alex tended to give orders like

© Springer Nature Switzerland AG 2019
L. Erdős, *Green Heroes*, https://doi.org/10.1007/978-3-030-31806-2_14

'a feathery Napoleon.' Yet Alex was a charming and loveable bird. He liked to dance, had a good sense of humour, and was a trustworthy colleague.

Alex died unexpectedly in 2007. He was probably the world's most famous bird. Obituaries were published in journals such as *The Economist*, *The New York Times*, and *Nature*. Alex will be remembered because he, in Pepperberg's words 'showed us just what a bird brain is capable of.'[1] Perhaps the phrase 'birdbrain' should not be taken as an insult any more.

Besides its enormous scientific importance, Pepperberg's pioneering study also has an ethical dimension. The high intelligence and rich emotional life of parrots (and probably many other animals) challenges the way these animals are usually kept in our homes. Confining parrots in small cages for their entire lives, where they have absolutely nothing to do, is a form of animal cruelty, irrespective of whether or not the bird can talk. Everyone who decides to keep parrots or other pets has to ensure that they have company and can engage in meaningful activities in an enriched environment.

Irene Pepperberg's studies continue with the grey parrots Athena and Griffin. The Alex Foundation, established by Pepperberg, not only supports these studies, but also works to enrich the lives of captive and companion animals and to protect wild populations, with a particular emphasis on parrots.

The extraordinary intelligence of dolphins and whales is widely known. In the studies conducted at the Marine Mammal Laboratory in Hawaii, the bottle-nosed dolphins Akeakamai and Phoenix learned an artificial language consisting of gestures and sounds. They were able to understand individual signals that were equivalent to words. The dolphins' vocabulary included words referring to objects, actions, relationships, and modifiers. Moreover, they could comprehend combinations of signals that were arranged according to simple grammatical rules; these combinations can be seen as equivalent to sentences. Akeakamai and Phoenix demonstrated dolphins' ability to generalise and form abstract concepts. Paradoxically, the results of these dolphin studies underline the necessity of terminating all research on captive cetaceans. Capturing dolphins and confining them to tanks for research or entertainment is morally unacceptable, as not even the largest aquaria can possibly meet the needs of these smart and emotional creatures.

Cetaceans are so different from us that it is hard to assess what they are capable of. What is sure is that they have huge brains, at least some of them recognise themselves in the mirror, have an excellent memory, and call one another by name. Whale populations seem to have their own cultures: whale songs have several dialects within individual species, and some tunes spread across the oceans much like human pop songs do. Dolphins have been observed to aid injured conspecifics or even humans. Both dolphins and whales use advanced acoustic communication systems, which have not yet been cracked by experts. It seems certain that cetaceans qualify as persons.

---

[1] Quoted in Lehrer, J.: Eggheads: How Bird Brains Are Shaking Up Science. The Boston Globe, September the 16th, 2007. http://archive.boston.com/news/globe/ideas/articles/2007/09/16/eggheads/

Remarkable mental abilities have also been demonstrated by sea lions. Among them, Rio and Rocky have probably been the most famous. They learned the meanings of gestural signs and even combinations of signs (equivalent to simple sentences). It became clear that the classification of objects into abstract categories is no problem for a sea lion. Today we know that the performance of sea lions is in some respects very similar to that of dolphins and apes. Many a volunteer participating in intelligence tests with Rio and Rocky felt somewhat embarrassed to be easily outperformed by sea lions.

Several people still believe that humans are the most intelligent beings on this planet. But this is an illusion. Intelligence is not a one-dimensional feature: there are several aspects, indeed, countless types of intelligences. Humans perform excellently in some types, but in other types we are easily outperformed by a whole range of non-human animals. To measure non-human animals by human standards is as mistaken as it would be to measure all sportspeople by the number of goals they have scored – which would be quite unfair to goalkeepers, not to mention sprinters or canoeists. Each animal's intelligence fits its environment, but neither is 'lower' or 'higher'; they are just different. As Frans de Waal wrote, 'Instead of turning the study of cognition into a contest, we should avoid putting apples next to oranges.'[2]

Of course, in a way humans are unique, just as every other species is unique. But by no means is the human species superior to the millions of other existing life forms.

Worth Reading

Pepperberg, I. M. (2009). *Alex & me*. New York: Harper.

Worth Browsing

alexfoundation.org

Worth Watching

Life with Alex: A memoir (2011)
Dolphins: Deep thinkers? (2003)
Extraordinary animals – The smartest sea lion (2007)

---

[2] de Waal, F. (2016): Are We Smart Enought to Know How Smart Animals Are? Granta Books, London, p. 248.

In the Front Line of Animal Advocacy – From Brigitte Bardot to Lek Chailert

Once a battle of a passionate few, the fight for the welfare and rights of animals has developed into a mainstream movement by the twenty-first century. Today the cause unites people with the most different backgrounds from all over the world. This chapter honours the achievements of four extraordinary women who have been working tirelessly to help animals in need.

Brigitte Bardot was the ultimate beauty and sex symbol of the 1960s. But in 1973, at the height of her career as a movie icon, she left the show business and started a completely new life as an animal advocate. She announced her retirement from films before her 39th birthday, and has been working for animals ever since.

But even before the turn, Bardot had been involved in animal advocacy. When she was filming on location, she would rescue stray dogs and cats. In 1962 she was the first celebrity in France to speak out against the cruel killing methods applied in slaughterhouses. She advocated the use of painless techniques. Her campaign was partly successful, as there were some legislative improvements. However, profit always comes first: regulations in favour of animals are designed so that they never threaten business interests. Therefore, industrial animal farming can only be defeated by eliminating consumer demand. Brigitte Bardot, herself a vegetarian for decades, encourages people not to eat meat, or at least avoid products from factory farms.

Once Bardot quit the world of cinema, she could devote all her time and energy to helping animals. Her decision to join the movement against the annual seal hunt in Canada received much publicity. She appeared at demonstrations and negotiated with politicians. In 1977, despite her extreme fear of travelling by plane, she flew to the ice fields of Canada. She embraced a baby seal and promised to do everything she can to end the slaughter. The visit was a media sensation. The photo showing Bardot and the seal pup is credited with leading to the European Community banning the importation of seal fur. Brigitte Bardot has been working hard ever since to keep the promise she made to the seal. 'I want to see this barbaric massacre stopped

© Springer Nature Switzerland AG 2019
L. Erdős, *Green Heroes*, https://doi.org/10.1007/978-3-030-31806-2_15

before I die,'[1] she said at a press conference. As yet she has not been able to melt the hearts of the sealers or the Canadian decision makers.

In 1986 Bardot launched The Brigitte Bardot Foundation to promote the rights and welfare of animals (her previous attempt to set up a foundation had failed a decade earlier due to her inexperience in organisational matters and bureaucracy). To secure the financial background of the new Foundation, she auctioned off her jewellery, personal belongings, books, and art collection. She even donated her home to the Foundation.

Bardot's show entitled *S.O.S. Animaux* ran on French television from 1989 to 1992, educating a broad audience about the various cruelties people inflict on animals. Her 1989 campaign against the massacre of elephants was instrumental in ivory products becoming unfashionable, naff, or even embarrassing objects. It was Bardot who called attention to the fact that each year tens of thousands of dogs are stolen in France to be sold to laboratories. Partly as a result of her warning, a whole network of thieves, dealers, and buyers was brought to court.

One of the primary aims of The Brigitte Bardot Foundation is to reduce the number of stray animals. They operate shelters and carry out sterilisation projects in France and support similar programmes in Eastern Europe, Central and South America, and Asia.

Brigitte Bardot and Paul Watson of the Sea Shepherd Conservation Society have a long history of co-operation, which goes back to the 1977 anti-sealing campaign. The Sea Shepherd named one of their vessels in honour of the French animal advocate. The respect is mutual. When, in 2012, Paul Watson was arrested because he had tried to stop an illegal shark-finning ship, Bardot offered to take his place in prison.

In 2016 Brigitte Bardot and her Foundation teamed up with Pamela Anderson to end the force-feeding of geese and ducks, a method used during the production of foie gras. Foie gras is the abnormally large, swollen liver of the birds. Though it is a luxury product no one really needs, some gourmets think its taste is worth the fear, misery, and pain of tens of millions of birds each year. Bardot thinks otherwise, so she tries to persuade people not to buy this product.

Bardot is an ardent opponent of bullfights and rodeos, heavily criticises keeping dolphins in aquaria, and believes that wearing fur is deeply unethical. 'Never forget that wearing a fur is wearing a cemetery on your back,'[2] she insists.

Brigitte Bardot is usually criticised because some think her biting remarks on certain cruel rituals or traditions border on racism. These critics, however, miss the point. Bardot denounces every form of cruelty to animals, no matter which race or religious group the perpetrator belongs to. This is why she has repeatedly criticised her fellow French nationals for eating horse meat or shooting pigeons. According to

---

[1] Quoted in: Bardot Cries for End of Seal Hunt. The Guardian, March the 23rd, 2006.

[2] Quoted in Jones, L. (2014): Brigitte Bardot: 'I've been a victim of my image.' Mail Online, November the 2nd, 2014. https://www.dailymail.co.uk/home/you/article-2815676/Brigitte-Bardot-ve-victim-image.html

Bardot, neither religious beliefs nor traditions can serve as excuses for torturing animals.

In several cases Bardot personally rescues abused animals and urges that the perpetrators should be brought to account for animal cruelty. Well in her eighties, she is still full of work. In addition to being active at her foundation, she also takes care of about one hundred rescued animals at her home near Saint-Tropez. 'I don't have time to waste thinking about ageing, because I live only for my·cause,'[3] she says.

Carol J. Adams is a feminist and an activist working for social justice. For her, social justice includes justice for non-human animals. 'Animal activists know that animals are like human beings because human beings *are* animals,'[4] she holds.

A life-changing event in Carol's life happened when her pony died. That evening, while biting into a hamburger, she realised that mourning the death of her beloved pony and eating the flesh of a slaughtered cow is a contradiction. She decided to abandon meat. Soon after this episode she also became aware that feminism and vegetarianism are inextricably linked. She explored the topic in her influential work, the 1990 classic *The Sexual Politics of Meat*. The book describes how feminism and vegetarianism, and patriarchy and meat eating are connected. Adams argues that the abuse of animals is related to the abuse of women. Consequently, feminism and vegetarianism are natural allies: opposing the exploitation of animals should go hand in hand with opposing the exploitation of women. In *The Sexual Politics of Meat* Adams also introduces the concept of 'the absent referent.' She points out that each meal containing meat necessarily involves an act of violence. However, 'to protect the conscience of the meat eater,'[5] the violence is not taken into account: the animal is mentally separated from the end product. Those who eat meat want to enjoy their meal, and for this, they have to pretend that the animals are absent. The fact that the meat was once a living and sentient animal is practically forgotten.

Carol J. Adams is a prolific author. In addition to her books about feminism and vegetarianism, she has published numerous other works, including practical guides for vegetarians and spiritual books for animal lovers. Adams is not only a theorist, but also an activist: she has been involved in several social causes and usually delivers lectures about animal advocacy across the world. The following words may be understood as her motto: 'Justice should not be so fragile a commodity that it cannot be extended beyond the species barrier of *Homo sapiens*.'[6]

Joyce Tischler is an attorney. Her clients are animals. No wonder she is usually referred to as 'the mother of animal law.'

---

[3] Quoted in: www.bardotbrigitte.com/brigitte-bardot-bio

[4] Carol J. Adams in an interview. Green, E. V.: Fifteen Questions For Carol J. Adams. The Crimson, October the 2nd, 2003. https://www.thecrimson.com/article/2003/10/2/fifteen-questions-for-carol-j-adams/

[5] caroljadams.com/spom-the-book

[6] Adams, C. J. (2010): The Sexual Politics of Meat: A Feminist-Vegetarian Critical Theory. Continuum, New York, p. 23.

As a young girl she adopted animals who were injured or abandoned. She decided to study law but her interest in helping animals did not fade. As a young lawyer she volunteered for the Fund for Animals. It was here that Joyce met lawyer Laurence Kessenick. In 1979, after recruiting other lawyers with similar attitudes to animals, they set up the organisation Attorneys for Animal Rights, changing its name to Animal Legal Defense Fund in 1984.

In 1981 Joyce Tischler was informed that the US Navy had killed hundreds of feral burros at one of its bases. Moreover, it turned out that there was a plan to kill another 5000 of these animals. Immediately starting a legal process and a series of negotiations, Tischler managed to save the burros from slaughter. The success attracted considerable attention and an increasing circle of law professionals became involved in the field of animal law.

In 2012, the Animal Legal Defense Fund and PETA filed suit against a roadside zoo in North Carolina. The zoo, which had been notorious for violating animal welfare regulations, had been keeping Ben the bear in a barren cage for years, under substandard conditions. Following the decision of the court, Ben was rescued and transported by air to his new home, a spacious and humane sanctuary. FedEx provided the flight for free; the plane used for the mission was called 'Bear Force One.' In a similar way, numerous other bears, horses, dogs, chimpanzees, and other animals have been rescued by ALDF.

The Animal Legal Defense Fund's mission is 'to protect the lives and advance the interests of animals through the legal system.'[7] It has been supporting animal advocacy in three major ways. First, the organisation has been involved in prosecuting perpetrators in animal cruelty cases. Second, it has supported efforts to enhance the legal protection of animals. Last but not least, the Fund has introduced animal law courses into several law schools in the US.

Sangdeaun Lek Chailert is 'the elephant whisperer of Thailand.' During her childhood she tried to help all kinds of animals, but she became especially interested in elephants.

The Asian elephant is Thailand's symbol yet this noble animal often has to endure barbaric acts of violence. Most elephants tourists meet in the country were captured in the wild as babies and trained to work in the logging industry or perform tricks for a paying audience. The training involves breaking the spirit of the elephant in an extremely cruel way.

Lek Chailert decided to help as many elephants as she possibly can. In 1995 she co-founded Elephant Nature Park, where rescued animals can live in peace and harmony in a large forested area. Most pachyderms have serious physical problems and psychological traumas when they arrive, but Chailert and her devoted team heal the elephants' bodies and souls. Visitors are allowed to enter the park and observe the elephants, but the animals are not forced to do any tricks and there are no elephant rides. The elephants are free to roam large areas in the jungle, and are treated with love and respect. In addition, the Elephant Nature Park is supporting analogous

---

[7] aldf.org/about-us/

projects throughout Thailand. The aim is to enable visitors to interact with elephants and observe their natural behaviour. The projects are able to generate income for the local inhabitants without the exploitation of the animals. This kind of ecotourism benefits elephants, the ecosystems they live in, and the rural Thai communities simultaneously. Chailert's approach has spread to Cambodia, and is also emerging in Myanmar.

During the catastrophic 2011 floods in Bangkok, volunteers of the Elephant Nature Park rented boats and rescued hundreds of dogs. Before long a shelter with a small hospital was built for them. This was how the Elephant Nature Park's dog project began. The shelter started to admit stray, neglected and abused dogs, as well as dogs rescued from the illegal dog meat trade. In addition, cats, buffaloes, and birds were also adopted.

Lek Chailert is not only interested in the welfare of elephants, but also in conserving their habitats. That is why her organisation, the Save the Elephant Foundation takes part in protecting existing forests and restoring degraded areas by planting trees. These efforts are supported by Buddhist monks in a unique way. They donate their orange robes to the conservationists, who tie the robes around trees. As these clothes are held in the highest esteem in Thai culture, most loggers do not dare to harm the 'protected' trees.

In 2005 Lek Chailert was selected by *Time* magazine as one of Asia's heroes. In 2010 she was named Woman Hero of Global Conservation, and in 2018 she won the Responsible Thailand Award in the category Animal Welfare. But not everyone was happy about her efforts. Those who wanted to continue the ruthless exploitation of elephants regarded Lek Chailert's work as a threat. There was a time when she had to endure repeated harassments, received death threats, and one of her elephants was poisoned. Today, however, her approach is gaining momentum in Thailand and the neighbouring countries. The Elephant Nature Park serves as a model. Lek Chailert has shown that working with elephants can be based on mutual respect and love.

Worth Reading

Adams, C. J. (2001). *Living among meat eaters: The vegetarian's survival handbook*. New York: Three Rivers Press.

Adams, C. J. (2005). *God listens to your love: Prayers for living with animal friends*. Cleveland: Pilgrim Press.

Adams, C. J. (2015). *The sexual politics of meat: A feminist-vegetarian critical theory*. New York: Bloomsbury.

Adams, C. J., & Breitman, P. (2008). *How to eat like a vegetarian even if you never want to be one: More than 250 shortcuts, strategies, and simple solutions*. New York: Lantern Books.

Bardot, B., & Huprelle, A.-C. (2019). *Tears of battle: An animal rights memoir*. New York: Arcade Publishing.

Craker, L. (2014). *The last elephant: The fight to save the elephants of Thailand*. Scotts Valley: CreateSpace.

Worth Browsing

aldf.org
caroljadams.com
www.saveelephant.org
loveandbananas.com
www.fondationbrigittebardot.fr

Worth Watching

The Age of Beasts (2017)
The Last Elephants in Thailand (2009)
Love & Bananas (2018)
A Cow at My Table (1998)
Vanishing Giants (2002)
How I Became an Elephant (2012)

Part II
Heroes for Nature

Beginnings of the Conservation Movement

In the course of western civilisation, wilderness, with very few exceptions, was regarded as worthless and hostile. Nature had to be destroyed, subdued, tamed or exploited. This attitude started to change fundamentally in the nineteenth century. The era of Romanticism marked the dawn of the conservation movement. More and more painters, poets, writers, and philosophers recognised the beauty of nature. At the same time it became evident that wilderness, which had seemed almost infinite previously, was shrinking rapidly and the last remnants of it deserved protection from human encroachment.

American poet and philosopher Ralph Waldo Emerson was among the first to notice the aesthetic and spiritual value of nature. His 1836 essay *Nature* is a central piece of transcendentalism, a world view stating that God suffuses nature, thus the divine can be experienced in nature.

The thinking of Emerson influenced a great number of philosophers, including his friend and mentee Henry David Thoreau. Thoreau was born in Concord, Massachusetts, in 1817. He graduated from Harvard College and worked as a teacher for a couple of years. Thoreau became interested in simple living. In 1845, he built a cabin in a forest at Walden Pond in the East of Massachusetts. He moved in on Independence Day, July the 4th, and spent there two years, two months, and two days. 'I went to the woods because I wished to live deliberately, to front only the essential facts of life, and see if I could not learn what it had to teach, and not, when I came to die, discover that I had not lived,'[1] he explained why he wanted to live in nature. During his stay in the woods he spent time cultivating a small bean-field, swimming, boating, visiting the nearby settlement, and, most importantly, going on outings. 'I frequently tramped eight or ten miles through the deepest snow to keep an appointment with a beech-tree, or a yellow-birch, or an old acquaintance among the pines,'[2] he wrote in *Walden*, a book about his experiences at the Pond.

---

[1] Thoreau, H. D. (2009): Walden or, Life in the Woods. Cosimo Classics, New York, p. 59.

[2] Thoreau, H. D. (2009): Walden or, Life in the Woods. Cosimo Classics, New York, p. 171.

© Springer Nature Switzerland AG 2019
L. Erdős, *Green Heroes*, https://doi.org/10.1007/978-3-030-31806-2_16

Thoreau was completely enchanted by the squirrels, the owls, the rainbow, the colours of the landscapes, and all of nature's inhabitants and phenomena.

While at Walden Pond, Thoreau completed a first draft of *A Week on the Concord and Merrimack Rivers*, an account of an excursion he had made with his brother a few years earlier. The book was published in 1849 but did not sell well: more than 700 unsold copies were returned to the writer. 'I have now a library of nearly 900 volumes, over 700 of which I wrote myself,'[3] Thoreau wrote in his diary.

Thoreau opposed the Mexican-American war and refused to pay poll tax because he thought the money would be spent on the war. He was arrested and had to spend one night in jail but was released the next day when someone paid his debt. He explained his way of resistance in his 1849 essay *Resistance to Civil Government* (later reprinted under the title *Civil Disobedience*), which had a profound influence on civil rights activists such as Mahatma Gandhi and Martin Luther King Jr.

Thoreau kept on visiting natural areas for shorter periods and he wrote extensively about his experiences and thoughts. He warned that the overexploitation of natural resources would have dire consequences and advocated the formation of protected areas and the wise management of private lands.

Thoreau's untimely death in 1862, at the age of just 44, cut short a remarkable career. Today Thoreau is widely regarded as a forerunner of modern conservation and a pioneer of the civil rights movement.

The romantic-transcendentalist views of Emerson and Thoreau culminated in the work of John Muir, the father of the national parks, one of the greatest icons in the history of nature conservation.

Muir was born in Dunbar, Scotland, in 1838. He had a quite harsh childhood due to the dour personality of his father, who regularly beat his children and discouraged them from leaving their home or playing with pals. In 1849 Muir's family moved to the US, settling in Wisconsin. John had to work on the farm but used his little free time to wander the countryside. Aged 22 he enrolled at the University of Wisconsin, where he attended courses on chemistry, geology, and botany, but never graduated. He became fascinated with nature and frequently went on outings, exploring the plant life of the woods. Muir took up a job in a factory, where he almost lost an eye in an industrial accident in 1867. The event changed his life completely. Muir decided that he would devote his life to studying the natural world. In September that year he set off from Indianapolis and walked some 1600 km to the Gulf of Mexico. 'My plan was simply to push on in a general southward direction by the wildest, leafiest, and least trodden way I could find, promising the greatest extent of virgin forest,'[4] he wrote in *A Thousand-Mile Walk to the Gulf*. He made the tour with little more than the clothes he was wearing. He often slept

---

[3] Quoted in Meltzer M. (2007): Henry David Thoreau: A Biography. Twenty-First Century Books, Minneapolis, p. 98.

[4] Muir, J. (1916): A Thousand-Mile Walk to the Gulf. Houghton Mifflin Company, Boston, pp. 1–2.

on the ground under the stars and at times hardly had anything to eat. Muir kept a journal and made sketches of the landscapes, the flora and the fauna, and contemplated man's place in nature.

From Florida Muir sailed to Cuba, and spent some time in Havana. As he could not find a ship bound for South America, his original destination, he decided to travel to California and visit the famous Yosemite Valley. Muir was reborn in Yosemite. He built a cabin, worked as a shepherd, climbed mountains, toured the Valley, and studied its geography. Muir was walking in nature even when everyone else was seeking shelter: at night, in snowstorms and hailstorms. During the 1872 earthquake he felt fortunate to observe this unique natural event firsthand. After some years Muir returned to society, married and started a family. He managed an orchard but regularly returned to Yosemite.

Muir gained fame through his studies and publications about the Yosemite and other natural areas. He was the first to suggest that the Valley had been formed by glaciers rather than an earthquake. In 1889 he guided his friend Robert Underwood Johnson in the area. The two thought that the Yosemite should be protected as a national park. Their proposal was supported by the Congress, creating Yosemite National Park in 1890.

Hardly had the Yosemite National Park been formed when, in the early 1900s, it was proposed that the Hetch Hetchy Valley within the National Park be flooded to supply San Francisco with drinking water. Muir opposed the project. A countrywide debate and a long legal process ensued, which is considered the first large-scale grassroots campaign in the conservation movement. Eventually the cause was lost when the building of the dam was authorised in 1913. But the conservationists' efforts were not in vain, as millions of citizens across the country became aware of the values represented by national parks and wilderness areas.

Resource Conservation vs. Preservation

The Hetch Hetchy case perfectly illustrates the differences between the two directions of the conservation movement in the nineteenth century. Resource conservation emphasised the instrumental value of nature, advocating the optimal use of natural resources for the benefit of present and future human generations. Led by Gifford Pinchot, this direction supported the flooding of Hetch Hetchy Valley. The other approach was preservation, which favoured setting aside large natural areas where any major human activity should be prevented. Preservationists, including John Muir, maintained that nature has intrinsic value, i.e., that nature is valuable in itself, independent of its benefits to humans. Today both resource conservation and preservation are placed under the umbrella of conservation.

In 1892 Muir co-founded the Sierra Club, of which he was the first president. The Sierra Club has led several successful campaigns for the protection and enlargement of existing national parks and the formation of new ones. Today the Sierra Club has more than 3.5 million members and is one of the most influential conservation organisations in the US.

It was Muir who proposed that a bureau be formed to protect and manage the national parks of the US. The National Park Service, set up in 1916, two years after Muir's death, clearly reflects Muir's ideas.

Muir's views about Native Americans were rather ambivalent. He greatly respected their harmony with nature, but occasionally, influenced by his upbringing and the dominant views of the time, he made unfortunate remarks on them. However, charges stating that Muir was a racist are completely mistaken. On the contrary: Muir strongly opposed the unjust treatment of Native Americans.

Muir's influence as a nature writer, philosopher, and conservation activist is immense. He led millions to appreciate nature's treasures and advocate conservation efforts. He played a prime role in the creation of some of America's most famous national parks. Muir wrote about the spiritual, aesthetic, and recreational value of nature. What was even more radical at his time, he stressed the intrinsic value of non-human beings and nature as a whole. 'No dogma taught by the present civilization seems to form so insuperable an obstacle in the way of a right understanding of the relations which culture sustains to wildness as that which regards the world as made especially for the uses of man,'[5] he wrote.

Probably there are more conservationists today than ever before, but the pressure on nature is also larger than is has ever been. The conservation movement has a long way to go before the destruction of natural values is halted. As Muir said, 'The battle for conservation will go on endlessly. It is part of the universal warfare between right and wrong. Fortunately wrong cannot last, soon or late it must fall back home to Hades, while some compensating good must surely follow.'[6]

Worth Reading

Muir, J. (1998). *A thousand-mile walk to the Gulf*. Boston: Mariner Books.
Muir, J. (2011). *My first summer in the Sierra*. Boston: Mariner Books.
Muir, J. (2018). *The Yosemite*. Mineola: Dover Publications.
Thoreau, H. D. (2009). *Walden or, life in the woods*. New York: Cosimo Classics.

---

[5]Quoted in: Sandler R. (2010) Environmental Virtue Ethics. In Keller, D. R. (ed.) Environmental Ethics: The Big Questions. Wiley-Blackwell, Chichester, pp. 252–256.

[6]www.sierraclub.org/rhode-island/about-us

Worth Browsing

thoreausociety.org
www.walden.org
www.sierraclub.org

Worth Watching

The Unruly Mystic: John Muir (2017)
National Parks: America's Best Idea (2009)

Aldo Leopold and the Land Ethic

From the nineteenth century onwards an increasing circle of notable philosophers and authors put forward the view that nature is intrinsically valuable, i.e., that nature's value is independent of its benefits to humans. This realisation gave rise to the emergence of nature-centred ethical systems focusing on ecological communities or ecosystems. The first of these ecocentric approaches was outlined by Aldo Leopold.

Aldo Leopold was born in 1887 in Burlington, Iowa. He grew up as an amateur naturalist. His father practised hunting but was proud of always following a strict hunting ethic. The young Aldo Leopold often accompanied his father on his hunting trips, but was also interested in other types of outdoor activities, including bird watching. He decided to pursue a career in forestry and graduated from Yale University as a forester in 1909. He worked for the US Forest Service in Arizona and New Mexico. His job was about the management of public lands. But he realised that some natural areas should be set aside without any management at all. He suggested that a wilderness area be designated in the Gila National Forest in New Mexico. After a few years, his proposal was accepted and the first wilderness area of the world was established, serving as a model for the creation of other wilderness areas.

As a forester Leopold took part in the eradication of great predators such as wolves, bears, and mountain lions, as these animals were regarded as pests. Leopold's attitude changed when he shot a wolf and watched her die. 'I was young then, and full of trigger-itch; I thought that because fewer wolves meant more deer, that no wolves would mean hunters' paradise,' he wrote. 'But after seeing the green fire die, I sensed that neither the wolf nor the mountain agreed with such a view.'[1] Lepold came to understand that large predators have a very important role in nature and they are essential for the health of the ecosystems.

---

[1] Leopold, A. (1986): A Sand County Almanac with Essays on Conservation from Round River. Ballantine Books, New York, pp. 138–139.

© Springer Nature Switzerland AG 2019
L. Erdős, *Green Heroes*, https://doi.org/10.1007/978-3-030-31806-2_17

In 1924 Leopold moved to Madison, Wisconsin, where he became interested in how conservation goals could be incorporated in the management of privately owned lands. He interviewed farmers and explored ways of sustainable agriculture. In 1933 Leopold published *Game Management*, the world's first textbook in the field. In the same year he became professor of game management at the University of Wisconsin. In 1935 Leopold co-founded The Wilderness Society to save the unspoiled natural areas of the US.

In 1935 Leopold launched an unusual ecological experiment. He bought a highly degraded farm in an area that is known as the sand counties. With the help of his family he started to restore the native hardwood and prairie vegetation. Moreover, Leopold observed and documented the results of his project, which equipped him with a robust ecological knowledge. Simultaneously, partly at Leopold's urging, similar works were being carried out in the arboretum of the University of Wisconsin. These projects marked the emergence of a new field, which is now known as ecological restoration.

Based on his professional background in ecology and his interest in ethics, Leopold wrote a book about nature and humans' relationship to it. Hardly had the manuscript been accepted for publication when Leopold died of a heart attack while helping to fight a wildfire on a neighbour's farm. His book entitled *A Sand County Almanac* came out posthumously in 1949.

A Sand County Almanac combines literary quality with scientific accuracy and highly readable style. It provides the audience with basic ecological facts wrapped in the finest nature writing. Leopold stresses the ecological significance of the seemingly unimportant or harmful creatures, discusses how natural entities are interconnected with one another, and describes the negative consequences of human irresponsibility and short-sightedness. Leopold was way ahead of his time in realising the importance of environmental education and citizens' involvement in conservation.

However, the book's most lasting effect is due to its contribution to the field of conservation ethics. Leopold argues that ethics should not only deal with how humans have to treat other humans. Instead, ethics should be extended so as to regard ecological communities as morally considerable. We have to realise that we are members not only of human society, but also of a wider, an ecological community. 'The land ethic simply enlarges the boundaries of the community to include soils, waters, plants, and animals, or collectively: the land,'[2] he wrote, adding that 'a land ethic changes the role of *Homo sapiens* from conqueror of the land-community to plain member and citizen of it. It implies respect for his fellow-members, and also respect for the community as such.'[3]

Leopold's land ethic is ecocentric. It recognises the intrinsic value of ecological communities (or, using Leopold's term, biotic communities) such as forests or

[2]Leopold, A. (1986): A Sand County Almanac with Essays on Conservation from Round River. Ballantine Books, New York, p. 239.

[3]Leopold, A. (1986): A Sand County Almanac with Essays on Conservation from Round River. Ballantine Books, New York, p. 240.

grasslands. Leopold's famous maxim describes how our actions should be judged based on how they affect ecological communities: 'A thing is right when it tends to preserve the integrity, stability, and beauty of the biotic community. It is wrong when it tends otherwise.'[4] This is the best-known ecocentric 'golden rule' thus far produced.

Holistic and Individualistic Ethical Views

Western thought has been dominated by anthropocentrism for a very long time. Anthropocentrism states that only humans are morally considerable. This view has been challenged from several directions.

There are strong arguments supporting the notion that at least some non-human organisms should be regarded as morally considerable. Sentiocentrism maintains that sentience (the capacity for suffering) is the necessary and sufficient condition for belonging to the moral community. In practice this means that numerous animals are morally considerable. Biocentrism implies a further expansion of our moral horizon, stating that all living organisms are morally considerable. The focus of both sentiocentrism and biocentrism is on individual organisms, that is why these approaches are termed individualistic.

Ecocentric views go one step further and realise the moral considerability of ecological levels above the individual organisms. Aldo Leopold's land ethic is a prime example of an ecocentric ethic: its focus is on whole ecological (or biotic) communities rather than their individual members.

It is important to note here that ecocentric views are usually not intended to replace individualistic ethical systems. Rather, ecocentric and individualistic approaches complement each other. The moral considerability of individual living beings and their ecological communities is not mutually exclusive. Thus, ecocentrism is compatible with both sentiocentrism and biocentrism.

Aldo Leopold's *A Sand County Almanac* belongs to the most important books in the history of conservation. Leopold's land ethic is radical but clear. By and large it is able to serve as a valid guiding principle in current conservation and is an inspirational source for environmental philosophers, conservation biologists, and nature lovers. Thanks partly to Leopold and his followers, more and more people begin to understand that we are linked to the ecological communities of the Earth in myriad ways. This knowledge will hopefully result in the emergence of a new ecological conscience. As Leopold put it, 'When we see land as a community to which we belong, we may begin to use it with love and respect.'[5]

---

[4]Leopold, A. (1986): A Sand County Almanac with Essays on Conservation from Round River. Ballantine Books, New York, p. 262.

[5]Leopold, A. (1986): A Sand County Almanac with Essays on Conservation from Round River. Ballantine Books, New York, pp. xviii-xix.

Worth Reading

Leopold, A. (1986). *A Sand County Almanac with essays on conservation from Round River.* New York: Ballantine Books.

Worth Browsing

www.aldoleopold.org
www.wilderness.org

Worth Watching

Green Fire: Aldo Leopold and a Land Ethic for Our Time (2011)

Under the Banner of the Giant Panda

The giant panda certainly belongs to the cutest and most beloved animals in the world. Numbering less than 2000 individuals, the giant panda personifies all threatened species and symbolises the efforts of the conservation movement. It is, of course, no coincidence that the giant panda is depicted in one of the most recognisable and successful logos ever devised: the logo of the WWF, a leading global conservation organisation.

The history of WWF goes back to the 1960 African trip of the biologist Julian Huxley, who was shocked to see how ruthlessly animals were being hunted and how rapidly natural habitats were shrinking. Upon his return to England he published articles about the situation, which arouse the attention of businessman Victor Stolan. Stolan suggested that a global conservation organisation be established. They asked Max Nicholson, a famous ornithologist, to undertake the organisational work. Nicholson recruited other devoted conservationists, and World Wildlife Fund was launched in 1961 with the goal of supporting conservation efforts all over the world.

WWF's logo was designed by Gerald Watterson and Peter Scott, who were inspired by Chi Chi, a giant panda at the London Zoo. 'We wanted an animal that is beautiful, is endangered, and one loved by many people in the world for its appealing qualities. We also wanted an animal that had an impact in black and white to save money on printing costs,'[1] Peter Scott said.

Among the earliest projects of WWF were the organisation's contributions to the protection of Galápagos Islands' natural treasures, and the establishment of Doñana National Park in Spain. WWF was instrumental in bringing together several nations to sign the Ramsar Convention, an international agreement for the conservation of wetlands.

Some of WWF's best-known campaigns started in the 1970s, including the programmes to save tigers and rhinos. In 1975 WWF launched its Tropical Rainforest Campaign to create jungle reserves in Africa, Asia, and South America. Another

---

[1] wwf.panda.org/knowledge_hub/history/

L. Erdős, *Green Heroes*, https://doi.org/10.1007/978-3-030-31806-2_18

campaign, entitled 'The Seas Must Live,' aimed to help cetaceans, seals, and turtles by creating marine reserves.

In 1979 WWF signed an agreement with China about collaboration in conservation issues. It was a historic event: WWF became the first international conservation organisation to be allowed to work in China. Among the most urgent tasks was to stabilise panda populations. Intensive field surveys began to reveal the distribution of the species and the most important threatening factors. WWF and Chinese authorities have set up new nature reserves and introduced anti-poaching patrols. Ecological corridors have been formed to enable the migration of panda populations between isolated reserves. A captive breeding programme was started. WWF's community-based programmes aim to ensure that panda conservation will also support the livelihood of the local human population. Thanks to all these efforts, after decades of decline, panda numbers have recently begun to increase.

Flagship Species

Everyone loves giant pandas and other attractive species, many of which are national symbols. It is relatively easy to mobilise people to protect these charismatic species, while hardly anyone is concerned if a seemingly unimportant insect or a not-too-popular rodent goes extinct. In practical conservation flagship species such as pandas, koalas, polar bears, and gorillas are ambassadors for the natural world. People who admire these creatures may start to value less appealing species as well. Also, flagship species can help raise funds for conservation activities, which will be helpful for countless other species. For example, if we protect the habitat of the giant panda, other animals and plants inhabiting the same ecosystem will also benefit.

In 1986 the organisation's name was changed from World Wildlife Fund to World Wide Fund for Nature (the acronym remained unchanged), while the US and Canada branches retained the old name.

Whale conservation has been in the focus of WWF from its beginnings. When it became clear that the International Whaling Commission was unable to halt the decline of whale populations, conservation organisations, including WWF, started to recruit nonwhaling nations to the Commission. With these new votes, commercial whaling was banned in 1986. Another notable success was achieved in 1994, when, at the urging of WWF and other organisations, the Southern Ocean Whale Sanctuary was designated.

In the 1980s WWF's debt-for-nature initiative was accepted and put into practice. Originally proposed by Thomas Lovejoy, the scheme means that a portion of the debt of certain countries is forgiven provided that the debtors fulfil some conservation measures such as the creation of nature reserves.

In the 1990s the focus of WWF was considerably widened: while habitat and species conservation remained the primary mission, WWF became more involved in environmental protection. The fight against global climate change has been a primary issue ever since. WWF has been lobbying governments to take action against global warming, and at the same time the organisation has been encouraging individual citizens to reduce their personal pollutions. In addition, WWF has been successful in persuading major companies about the importance of corporate responsibility.

In 2007 WWF started Earth Hour, an event intended to raise global awareness of climate change. Originating in Sydney, Earth Hour rapidly developed into an international event, held towards the end of March each year. Today hundreds of millions of people and thousands of cities from virtually every country of the world participate by turning off non-essential electronic appliances. Between 8:30 p.m. and 9:30 p.m., households, offices, and tourist attractions around the globe go dark. What is probably even more important, many individuals and public institutions make resolutions to consume less energy during the whole year.

WWF regularly publishes reports to inform the general public and decision makers about conservation and environmental matters. Published every two years, the Living Planet Report is the organisation's most important publication, a comprehensive overview of the biosphere's health. As noted by the 2018 Report, 'We are the first generation that has a clear picture of the value of nature and the enormous impact we have on it. We may also be the last that can act to reverse this trend.'[2]

WWF is one of the most respected conservation organisations. It has millions of supporters throughout the world, and is working in more than 100 countries. The current activities of WWF are grouped into six major topics: forests; oceans; freshwater ecosystems; biodiversity and threatened species; food sustainability and food security; climate and energy. The mission of WWF is a mission for all of us: 'to build a future in which people live in harmony with nature.'[3]

Worth Browsing

wwf.org
wwf.panda.org
www.earthhour.org

---

[2] wwf.panda.org/knowledge_hub/all_publications/living_planet_report_2018/
[3] wwf.panda.org/our_ambition/

Arne Naess and the Deep Ecology Movement

With the rapid development of the science of ecology around the middle of the twentieth century, holistic views about life on Earth gained importance. These views recognised the value of ecological levels above the individual organisms. Aldo Leopold's land ethic stressed the moral considerability of ecosystems. This pioneering concept was followed by other ecocentric approaches, one of the most influential of which was the deep ecology platform of Arne Naess.

Arne Naess was born in 1912 in Slemdal, Norway. He started to be interested in philosophy as a secondary school student and graduated from the University of Oslo in 1933. After earning his doctorate, he spent a short period in the USA, where he studied animal behaviour. However, he felt so sorry for the rats that he abandoned the job. In 1939, although only 27 years old, he was appointed full professor at the University of Oslo. After the German invasion of Norway, Naess was active in the non-violent resistance to the occupation.

Since his teens Naess was passionate about mountaineering. From the 1930s he spent much time in the Hallingskarvet mountain region northwest of Oslo, working in his hut, going on outings, and enjoying the landscape. In 1950 he led the first expedition to climb Tirich Mir, a 7690 m high mountain peak in Pakistan. He was so much in awe of mountains that he always reminded his fellow climbers that mountains may be ascended but never conquered.

Naess turned his attention to environmental issues after reading Rachel Carson's influential book *Silent Spring*. He participated in Green Party politics and was the leader of the Norwegian branch of Greenpeace. In a 1970 demonstration he chained himself to rocks near a waterfall to protest against the building of a dam. Also, he regularly spoke out for conservation. But Naess' most important contribution to the green movement was his environmental philosophy.

Influenced by the views of Benedict Spinoza and Mahatma Gandhi, as well as Eastern thought, particularly Buddhism, Arne Naess developed his philosophy in the early 1970s. Naess distinguished between two types of environmental protection. Shallow ecology, in his understanding, applies technological solutions to mitigate ecological problems without revising our fundamental beliefs regarding our

© Springer Nature Switzerland AG 2019
L. Erdős, *Green Heroes*, https://doi.org/10.1007/978-3-030-31806-2_19

place in nature. The efforts of shallow ecology may bring some short-term improvements and are therefore welcome. However, the real solution of the ecological crisis requires a completely different approach, deep ecology, which challenges the misconception that humans are superior to nature.

Arne Naess called his personal ecological philosophy 'ecosophy T', where 'T' refers to Tvergastein, the name of his cabin in the Hallingskarvet range. According to the holistic vision of Naess, relations between entities are more important components of the world than entities themselves. Individual beings live in a complex network, connected to one another in many ways. The survival of organisms would not be possible without this intricate web of life. Thus, reality should not be conceived as consisting of separate and isolated entities. This applies to humans as well: we depend on other creatures and the whole of nature, which means that by harming nature we harm ourselves. In Naess' view we are so strongly connected to nature that our Self includes all other living beings, in fact, our Self is virtually identical to nature. In reality there are no differences among humans and other living organisms. As we are one with nature, protecting nature is best understood as self-defence. From this perspective, being an environmentalist does not mean having to make sacrifices. Quite on the contrary, it means living a happy and full life. Therefore, true Self-realisation can only be achieved by supporting the realisation and fulfilment of all other creatures. 'We seek what is best for ourselves, but through the extension of the self, our "own" best is also that of others,'[1] Naess wrote.

Naess emphasised that his personal ecological philosophy, 'ecosophy T' is but one version of deep ecology. He encouraged others to invent their own ecosophies. While individual versions of deep ecology are variable, there are some basic principles that are common to all of them. These form the platform of the deep ecology movement, as formulated by Arne Naess and George Sessions:

1. The flourishing of human and non-human life on Earth has intrinsic value. The value of non-human life-forms is independent of the usefulness these may have for narrow human purposes.
2. Richness and diversity of life-forms are values in themselves and contribute to the flourishing of human and non-human life on Earth.
3. Humans have no right to reduce this richness and diversity except to satisfy vital needs.
4. Present human interference with the non-human world is excessive, and the situation is rapidly worsening.
5. The flourishing of human life and cultures is compatible with a substantial decrease of the human population. The flourishing of non-human life requires such a decrease.
6. Significant change of life conditions for the better requires change in policies. These affect basic economic, technological, and ideological structures.

---

[1] Naess, A.; Rothenberg, D. (1989): *Ecology, Community and Lifestyle*. Cambridge University Press, Cambridge, p. 175.

7. The ideological change is mainly that of appreciating life quality (dwelling in situations of intrinsic value) rather than adhering to a high standard of living. There will be a profound awareness of the difference between big and great.
8. Those who subscribe to the foregoing points have an obligation directly or indirectly to participate in the attempt to implement the necessary changes.

The frugal lifestyle of Naess was in harmony with his ecological views. But he pointed out that simplicity is not equal to asceticism or unhappiness. Instead of spending a lot of money and buying a lot of goods he had a modest yet satisfied and joyful life.

Though primarily concerned with nature and biodiversity, the deep ecology platform and especially Naess' 'ecosophy T' are also strongly related to the animal rights movement as they recognise the inherent value of animals and, consequently, their right to a decent life. Naess insisted that non-vital human interests can never trump the vital interests of animals. Accordingly, he opposed animal experiments for cosmetic or military purposes.

Naess was respected not only because of his wisdom but also because of his sense of humour. His humble personality was almost legendary. On one occasion he was invited by philosopher Alan Drengson to visit an aikido class. Naess arrived there before his host, and was met by an assistant who did not recognise him. Naess was asked to help in preparing the room for the training session. And the 75 years old professor took a broom and started to sweep the floor without hesitation...

Naess died in 2009, at the age of 97. He is regarded as one of the most influential environmental thinkers of the twentieth and early twenty-first centuries. His thinking has been incorporated into Norwegian environmental education. He was extremely popular in Norway: a survey showed that he was the person Norwegian children and teenagers most wanted to meet. But his influence is not limited to his homeland. His book *Life's Philosophy* has been able to reach a wide audience beyond academic circles. His ideas have enriched the international green movement and continue to inspire environmental philosophers, conservation activists, and green politicians.

Naess was convinced that human advancement entails the widening of the self from a narrow and illusory ego to a real Self. Naess put it this way: 'Early in life, the social "self" is sufficiently developed so that we do not prefer to eat a big cake alone. We share the cake with our friends and nearest relations. We identify with these people sufficiently to see our joy in their joy, and see our disappointment in theirs. Now it is time to share with all life on our maltreated Earth through the deepening identification with life forms and the greater units, the ecosystems, and Gaia, the fabulous, old planet of ours.'[2]

---

[2] Naess, A. (1995): Self-Realization: An Ecological Approach to Being in the World. In Drengson A.; Inoue, Y. (eds.) The Deep Ecology Movement: An Introductory Anthology. North Atlantic Books, Berkeley, pp. 13–30.

Worth Reading

Devall, B., & Sessions, G. (2001). *Deep ecology: Living as if nature mattered*. Layton: Gibbs Smith.

Naess, A. (2008). *Life's philosophy: Reason and feeling in a deeper world*. Athens: The University of Georgia Press.

Naess, A., & Rothenberg, D. (1989). *Ecology, community and lifestyle*. Cambridge: Cambridge University Press.

Worth Watching

The Call of the Mountain (1997)

Living with Big Cats – The Story of Joy and George Adamson

On Christmas Eve 1942 Mr and Mrs Bally attended a party in Kenya. Peter Bally was a Swiss botanist. Her wife, whom he called Joy, was doing paintings on the wildlife of Africa. She had been born in Troppau, in the northern part of the Austro-Hungarian Empire (today Opava, in the Czech Republic), and had come to Kenya in 1937. Also present at the party was George Adamson, who had been living in Kenya since 1924. After trying various jobs, including road building, farming, gold prospecting, and safari guiding, Adamson had found the job that suited him perfectly: he had become a game warden.

Joy and George fell in love. Joy divorced Peter Bally and in 1944 she married George Adamson. So began a remarkable story of a remarkable couple, a story set in the Kenyan wilderness, a story that was as much international as it was African.

As a game warden, George Adamson was responsible for an area the size of Great Britain. He could rely on the support of rangers he had recruited. Some of them were converted poachers, who had an in-depth knowledge of the region and its wildlife. In 1956 George was forced to shoot a lioness charging one of his colleagues. When the animal was dead, Adamson heard cries from among the nearby rocks. Checking the crevices he found the three cubs of the lioness. He brought them to camp, where they were taken care of by Joy. The couple thought they could not cope with three lions, so they kept the smallest, whom they named Elsa, and sent the other two to the Rotterdam Zoo. Later, George Adamson thought it was a mistake not to keep all three cubs.

The Adamsons decided to return Elsa to the wild, which was a pioneering endeavour. Thanks to the training sessions devised by the Adamsons, Elsa learned how to survive in the wild on her own. She even had a litter of cubs. Elsa became a wild lion, but maintained her relationships with humans and regularly returned to the Adamsons' camp. Some of the Adamsons' guests were quite surprised how friendly Elsa was. In 1960 David Attenborough visited the Adamsons to make a film about Elsa. He went to bed after the long and exhausting journey. He was woken up by Elsa, who was lying on him…

© Springer Nature Switzerland AG 2019
L. Erdős, *Green Heroes*, https://doi.org/10.1007/978-3-030-31806-2_20

Unfortunately, Elsa contracted a tick-borne disease and by the time the medicine arrived at the remote camp it was already too late. Elsa the lioness died in 1961. The Adamsons had to take care of Elsa's cubs, who were eventually released into the Serengeti National Park.

Joy and George documented the events meticulously in their diaries, made a huge number of photos and filmed some of the most important happenings. *Born Free*, Joy Adamson's account of the extraordinary story was based on these sources. The book received favourable critics and became an international bestseller. Two sequels followed: *Living Free* and *Forever Free*. Joy donated the proceeds from the books to the conservation of African wildlife.

Elsa and her foster parents were so popular that the story had to be told in movie form. Starring Virginia McKenna and Bill Travers, *Born Free* came out in 1966 and became a box office hit. George served as adviser on the lions, while Joy checked that the animals were treated humanely. A total of 24 lions were used during the shooting of the film. Some of them came from circuses, others arrived from private collections or were orphaned cubs. Once the filming was over, the Adamsons and the stars who played them in the movie wanted to save and rewild all lions. Unfortunately, most of the animals were sent to zoos. However, they were allowed to rehabilitate three lions.

The Adamsons devoted their lives to returning big cats to the wild. After the rehabilitation of the initial few lions, Joy Adamson turned her attention to cheetahs and leopards, while George Adamson continued his work with lions. This was a parting of the ways for them: they started to live in separate locations. However, they did not divorce, remained on friendly terms, and respected each other's work.

When Joy Adamson received a female cheetah who had been kept as a pet, she immediately started a rehabilitation project. Thanks to Joy's love and perseverance, the cheetah learned to hunt, became self-sufficient, and raised several cubs. The life of Pippa the cheetah is immortalised in Adamson's books *The Spotted Sphinx* and *Pippa's Challenge*.

In 1976 Joy Adamson acquired an orphaned leopard whom she named Penny. After some training Penny was gradually released into the wild in the Shaba Reserve; her adventures are described in the volume titled *Queen of Shaba*.

Besides working with her animals, writing books, and painting, Joy frequently delivered lectures virtually all over the world. She promoted conservation issues and raised money for her Elsa Conservation Trust. She was still very active when she was getting close to her 70th birthday.

On the evening of January the 3rd, 1980, Joy Adamson did not arrive back at camp from her usual evening walk. Her dead body was discovered some 200 m from the camp, lying on the road. At first it was thought she had been killed by a lion, but it soon turned out that she had been stabbed to death by one of her employees. The police caught the murderer, who was sentenced to life imprisonment.

Also known as Baba ya Simba (father of lions in Swahili), George Adamson was the friend and protector of lions. In addition to the three lions from the movie, he received numerous orphaned, rescued, and captive-bred lions. One of them was flown by KLM from Rotterdam to Africa free of charge. George Adamson's goal was to ensure that these animals have a chance to live in the natural environment they belong to.

In 1970 George Adamson settled at Kora to continue his rehabilitation programme in a mostly unspoiled environment. The camp, a few thatch-covered huts surrounded by a fence, was situated in the middle of nowhere, deep in the African bush, isolated from civilisation. George Adamson lived with his brother Terence, some African employees, and the lions. In 1971 Tony Fitzjohn joined the staff. He received no salary but enjoyed working with lions and was a huge asset to Adamson. They would sleep under the stars, get up before dawn, and spend time with the lions in the morning and the afternoon. 'The feeling of tranquillity and unity with nature that we experienced when we were out walking with the lions was a major part of why I loved my life so much,'[1] Fitzjohn wrote. After noon they used to have a siesta, as neither the lions nor the men wanted to move in the boiling hot weather. In the evenings they talked, read books, or listened to the radio.

The camp was known as Kampi ya Simba, i.e. Camp of the Lions. But, in addition to the lions, there were countless other interesting animals around and in Adamson's camp. Among the visitors were numerous agamas, birds, ground squirrels, porcupines, mongooses, and a civet cat, while scorpions and snakes belonged to the more frightening guests.

The introduction of lions to the wild was a risky job. Lions are potentially dangerous: working with them requires constant alertness. The tiniest human failure can have fatal consequences. When one of the lions wounded the child of the local park warden, Adamson had to move his camp to another location. In 1977 Adamson himself was seriously bitten by one of the lions, which was all the more problematic when taking into account that the nearest medical post was some 130 km away.

But it was not the lions who represented the greatest threat at Kora. The area was controlled by poachers and bandits, who decimated the wildlife and did not hesitate to kill humans. On August the 20th, 1989, near Adamson's camp, bandits stopped a car with two persons inside: a German woman who was visiting Adamson, and her security escort. The assailants fired some shots, dragged the man out of the car and broke his legs, and attacked the woman. Upon hearing the gunfire, Adamson grabbed his revolver, sprang into his car with his staff, and drove to the scene. They were met with a hail of bullets. Adamson and two of his employees were killed but the ensuing confusion allowed the German woman and her escort to escape. George Adamson was laid to final rest in Kora, near the grave of his brother and that of his beloved lion called Boy.

The Adamsons had to kill animals to support their felines, though they killed only as many as was absolutely necessary. Their approach remains controversial, but their results cannot be denied. They succeeded in reintroducing several big cats to the wild. Some of these animals were killed by man, some died of natural causes, but many survived and their descendants still live in Africa.

---

[1] Fitzjohn, T.; Bredin, M. (2011): Born Wild: The Extraordinary Story of One Man's Passion for Africa. Crown Publishers, New York, p. 29.

The indirect impact of the Adamsons' work is probably even more significant. Their exceptional relationships with animals became world-famous and changed people's attitudes towards great predators. After *Born Free* the animals that had previously been considered savage beasts started to be regarded as individuals having charismatic personalities, as unique characters worthy of protection. The Adamson's dedication raised public interest in the conservation of African wildlife. The incomes from the books and the movie were used to fund conservation projects, and both Joy and George supported anti-poaching patrols. Kora was declared a national park largely due to the activity of George Adamson. Virginia McKenna and Bill Travers were so impressed by the animals that they set up the Born Free Foundation, which works to protect and help both free-living and captive animals.

Today the populations of all three species of African big cats are decreasing. Lions, leopards, and cheetahs are threatened by habitat loss, hunting, poisoning, the decline of prey populations, the illegal trade in their body parts believed to be of medicinal value, and unregulated tourism. 'Destroying the wilderness, and robbing its prospects of peace and of game, man leaves only the promise of danger. He has killed ten of my lions and murdered my wife,'[2] George Adamson wrote.

Worth Reading

Adamson, J. (1969). *The spotted Sphinx*. London: Collins & Harvill.

Adamson, J. (1972a). *Pippa's challenge*. New York: Ballantine Books.

Adamson, J. (1972b). *Joy Adamson's Africa*. New York: Harcourt Brace Jovanovich.

Adamson, J. (1980). *Queen of Shaba: The story of an African leopard*. New York: Harcourt Brace Jovanovich.

Adamson, J. (1981). *Friends from the forest*. New York: Harcourt Brace Jovanovich.

Adamson, G. (1986). *My pride and joy: An autobiography*. London: Collins Harvill.

Adamson, J. (1990a). *Living free*. London: HarperCollins.

Adamson, J. (1990b). *Forever free*. London: HarperCollins.

Adamson, J. (2000). *Born free: A lioness of two worlds*. New York: Pantheon Books.

Fitzjohn, T., & Bredin, M. (2011). *Born wild: The extraordinary story of one man's passion for Africa*. New York: Crown Publishers.

Worth Browsing

www.bornfree.org.uk
www.elsamere.com

---

[2] Adamson, G. (1987): My Pride and Joy. Simon and Schuster, New York, p. 13.

Worth Watching

The Born Free Legacy (2010)
Elsa's Legacy: The Born Free Story (2011)
Lord of the Lions: Adamson of Africa (1989)
To Walk with Lions (1999)
Elsa: The Lioness That Changed the World (2010)
The Lions Are Free (1969)

The Modern Noah – Gerald Durrell's Mission to Save Endangered Animals

When the last individual of a species dies, the species is extinct. With each species that is lost, its unique evolutionary history spanning millions of years is ended once and for all. Many of us think humanity has no right to eradicate other species. And some do everything to prevent further losses and to bring back threatened species from the brink of extinction. Gerald Durrell was a pioneer in this field.

Gerald Durrell was born in 1925 in Jamshedpur, India, where his father worked as an engineer. When Gerry was three the family moved to England. Between 1935 and 1939 they lived in Corfu. The young boy was obsessed with the animals inhabiting the Mediterranean island. He made friends with Theodore Stephanides, a Greek polyhistor and naturalist, who became his mentor. 'My childhood in Corfu shaped my life,'[1] Durrell wrote in his book *My Family and Other Animals*, which is a humorous account of his early wildlife adventures. He spent most of his time observing animals – and taking a large number of various creatures home, from scorpions to toads to owls. His passion for collecting animals was not always welcome by the other family members. Nor can this boyhood pastime be regarded as compatible with the current understanding of animal advocacy and conservation. However, Gerald Durrell eventually found a vocation that fitted his love of animals and that was also useful for the animals themselves.

After the family returned to England, Durrell worked for a pet shop, then a farm, and later became an animal keeper at a zoo. However, he quickly became disillusioned with how the animals were treated. His dream was to found and operate his own zoo.

In 1947 Durrell organised his first animal-collecting expedition, using his inheritance from his father. During this and subsequent journeys to Africa and South America, he captured animals for British zoos. Following the suggestions of his wife Jacquie and his brother Lawrence, Gerald Durrell started to write books to raise money. His writings were so successful that he was able to finance new expeditions. The 1957 trip to West Africa was his first expedition on which he collected

---

[1] Durrell, G. (2004): My Family and Other Animals. Penguin Books, New York.

© Springer Nature Switzerland AG 2019
L. Erdős, *Green Heroes*, https://doi.org/10.1007/978-3-030-31806-2_21

animals for his own zoo. The problem was that the zoo only existed as a plan. Therefore, the animals had to be accommodated in the garden of Durrell's sister Margaret. Now that he had the animals, Durrell had to find a permanent place for his zoo. The task was more difficult than he had thought, but finally he was able to secure the perfect site: a seventeenth century manor house surrounded by a park on the Channel Island of Jersey. The zoo opened in 1959.

Durrell heavily criticised the zoos of his time because they were not interested in conservation, and did not treat the captive animals appropriately. Durrell insisted that the existence of any zoo is justified only if it meets three basic criteria: conservation, animal welfare, and education. First and foremost, the primary aim of the zoo should be to save endangered species. Durrell thought zoos should breed threatened animals so that if they go extinct in the wild there still will be a captive population. Once the threatening factors are eliminated, animals can be reintroduced into their original habitats. Durrell also pointed out that if zoos start to have breeding populations, there will be no need to capture animals from the wild. As for animal welfare, enclosures should mimic the natural environment of the species, and provide the animals with adequate space and places to hide from the view of visitors. Social species should be kept in pairs or groups. Environmental enrichment has to be deployed to keep animals active and avoid boredom. Overall, the design of the enclosures should enable the animals to express their natural behaviour. Durrell's third criterion was education. He emphasised that zoos must inspire the visitors to respect the animals and inform the public about the threats facing the natural world.

Gerald Durrell and Modern Zoos

Although his views on zoos were dismissed initially, today most modern zoos try to adhere to the guidelines outlined by Durrell. However, there are still some zoos and aquaria that do not take part in conservation efforts, have very poor animal welfare standards, and convey a harmful message to the public about how animals should be treated. We have to put pressure on them to change their policies. The simplest approach is to boycott these facilities. Only visit zoos that are accredited by independent organisations such as the Association of Zoos and Aquariums or the European Association of Zoos and Aquaria. In addition, animal shelters that rescue local wildlife may also be open to the general public; these 'zoos' are also worth visiting. However, never go to zoos that only serve human entertainment. Do not support aquaria and marine parks that keep dolphins, as even the largest tanks are too small for these animals.

In 1963 Durrell set up the Jersey Wildlife Preservation Trust to support his zoo and its conservation efforts (after the founder's death it was renamed Durrell Wildlife Conservation Trust). The Trust has had notable successes in captive breeding programmes. For example, they saved the pink pigeon, a bird endemic to Mauritius. In the mid-1970s the extinction of the species seemed inevitable as its total population was a mere ten individuals. However, Durrell started a captive breeding programme, the first of its kind in the world. Captive-bred pigeons have successfully been released into the wild, and today the population consists of some 400 birds. The species is still vulnerable but its prospects are much better than they were a few decades ago. Durrell always emphasised that, whenever possible, species have to be protected in their natural habitats. Captive breeding should be used only as a last resort if the species' survival in the wild is at high risk. But the final goal is always the reintroduction into the wild. To ensure the long-term survival of the reintroduced animals, their habitats have to be restored and protected. That is why the Durrell Wildlife Conservation Trust, in co-operation with local communities and organisations, is carrying out ecological restoration and habitat conservation projects. The focus is mainly on islands, as island ecosystems are particularly vulnerable to destructive human activities. The Trust also participates in cutting-edge research activities to make conservation more efficient, and runs a training programme for conservation practitioners from all over the world.

Durrell was working strenuously for rare species. But he was never satisfied with his achievements and was upset about the rate of destruction he had to witness around the globe. 'There would be a dreadful outcry if anyone suggested obliterating, say, the Tower of London, and quite rightly so; yet a unique and wonderful species of animal which has taken hundreds of thousands of years to develop to the stage we see today, can be snuffed out like a candle without more than a handful of people raising a finger or a voice in protest. So, until we consider animal life to be worthy of the consideration and reverence we bestow upon old books and pictures and historic monuments, there will always be the animal refugee living a precarious life on the edge of extermination, dependent for existence on the charity of a few human beings,'[2] Durrell said. His constant worries about endangered animals might have been one reason for his self-destructive lifestyle: he smoke too much, ate too much, and drank too much.

From the 1980s onwards Durrell was facing serious health problems. Aged 65, despite having undergone hip replacement operation, he led an expedition to Madagascar. Upon his return his health deteriorated further. Gerald Durrell died in 1995.

Durrell was a self-taught naturalist who became a leading conservationist. He introduced captive breeding as a conservation tool. He changed the way animals were kept in zoos. Last but not least, Durrell was a great educator, who succeeded in drawing people's attention to the wonders of the animal kingdom and the importance of every living being in the biosphere's intricate network. As he put it, 'The great ecosystems are like complex tapestries – a million complicated threads, interwoven, make up the whole picture. Nature can cope with small rents in the fabric; it can even, after a time, cope with major disasters like floods, fires, and earthquakes. What nature cannot cope with is the steady undermining of its fabric by the activities of man.'[3]

---

[2] Quoted in Zama, M. C. (ed.) (2004): Prose for Our Times. Orient Longman, Kolkata, p. 141.
[3] Quoted in Desmond, K. (2017): Planet Savers: 301 Extraordinary Environmentalists. Routledge, Abingdon, pp. 112–113.

Worth Reading

Durrell, G. (1994). *The Aye-Aye and I*. New York: Touchstone.
Durrell, G. (2004). *My family and other animals*. New York: Penguin Books.
Durrell, G. (2005). *A zoo in my luggage*. London: Penguin Books.
Durrell, G. (2007a). *Menagerie Manor*. London: Penguin Books.
Durrell, G. (2007b). *Golden bats and pink pigeons*. Chichester: Summersdale.
Durrell, G. (2011a). *The stationary ark*. London: Bello.
Durrell, G. (2011b). *Catch me a Colobus!* London: Bello.

Worth Browsing

www.durrell.org

Worth Watching

The Wild Life of Gerald Durrell (2005)
To the Island of the Aye-Aye (1991)
Ark on the Move (1982)
The Stationary Ark (1975)
Himself and Other Animals (1995)
Gerald Durrell: The Man Who Built the Ark (1995)

Farley Mowat Never Cried Wolf

'We are currently witnessing the most significant conflict ever to engage the human species,'[1] the celebrated Canadian author Farley Mowat wrote. A World War Two veteran, he did not mean a conflict between armies, but a conflict between those who are destroying the natural world and those who want to protect it. Mowat faced no dilemma about which side to join: '…only one course was open to me. I had to become a full-fledged member of the conspiracy to save the planet.'[2]

Farley Mowat was born in Belleville, Canada in 1921. His father worked as a librarian so Farley was surrounded by books during his childhood. Besides reading, the boy's other hobby was nature. From 1933 the family lived in Saskatoon. Farley wandered the prairie with his dog Mutt, kept squirrels, snakes, and owls as pets, and with some friends even formed a naturalists' club called the Beaver Club. Farley was a keen birdwatcher and aged 14 became Canada's youngest person to hold an official bird-bander's permit. In 1935 he joined his great-uncle on a birding expedition to the tundra near the Hudson Bay. For a short period he had a weekly column entitled *Birds of the Season* in the local newspaper.

During World War Two Mowat fought on the Sicilian and Italian front lines. To escape the brutality of war, he began writing stories about his childhood experiences and his beloved dog. More than a decade later, the stories evolved into the book *The Dog Who Wouldn't Be*. The war also changed the way he regarded hunting. In his younger years Mowat used to hunt birds, but during the war he experienced what it was like to be hunted. Upon his return to Canada, he gave up hunting forever.

In 1946 Farley Mowat enrolled at the University of Toronto. During his years as a student he spent as much time as he could in the barrenlands west of the Hudson Bay, where he studied wolves, caribou, and other wildlife. He became increasingly interested in the Ihalmiut people inhabiting the region. Mowat learned their language

---

[1] Mowat, F. (2011): Foreword. In: Hunter, E. (ed.) The Next Eco Warriors. Conari Press, San Francisco, p. xiii.

[2] In: Writers' Trust of Canada (2011): A Writer's Life: The Margaret Laurence Lectures, Emblem Editions, p. 147.

© Springer Nature Switzerland AG 2019
L. Erdős, *Green Heroes*, https://doi.org/10.1007/978-3-030-31806-2_22

and documented their culture. He admired them because they lived in harmony with their environment, respected nature, and took only as much as was necessary for their survival. Mowat became aware of the fact that the Ihalmiut were about to vanish. Their traditional way of life was being devoured by industrial civilisation while the populations of caribou, their most important prey, were being decimated by commercial hunting.

It was during his time with the Ihalmiut that Mowat decided to become a full-time writer. After graduating from university in 1949, he published articles about the wildlife and the native people of the Arctic region and started to write the book *People of the Deer*, the aim of which was to mobilise the public and the Canadian government to save the Ihalmiut as well as other Inuit peoples. The book came out in 1952 and was later followed by other books with a similar topic, including *The Desperate People* (1959) and *Walking on the Land* (2000).

Mowat felt a sense of kinship with the native peoples of the barrenlands, but he also felt sympathy for the non-human inhabitants of the region. Wolves usually exist in the popular imagination as bloodthirsty animals worthy of complete eradication. The hunting lobby have repeatedly tried to reinforce these erroneous beliefs with carefully built up propaganda. When the caribou populations dropped, blaming the wolves was easier than discussing the overexploitation of the wildlife. Wolves and caribous had been co-existing for millennia until indiscriminate hunting started to threaten both of them. Wolf seemed the perfect scapegoat. Based on his research on the wolves of the barrenlands, Mowat tried to enhance the wolves' reputation with *Never Cry Wolf*, probably his best and most famous book. Mowat described the wolves' important ecological role and denounced the wanton extermination of these wonderful creatures. 'Whenever and wherever men have engaged in the mindless slaughter of animals (including other men), they have often attempted to justify their acts by attributing the most vicious or revolting qualities to those they would destroy; and the less reason there is for the slaughter, the greater the campaign for vilification,'[3] Mowat wrote in his book. Despite the serious message, the book is highly entertaining and humorous, thanks to Mowat's extraordinary talent as a storyteller. *Never Cry Wolf*'s genre is subjective non-fiction; it is a mix of objective facts and invented portions. It gives a generally truthful picture of wolf behaviour but should not be understood as a scientific description. *Never Cry Wolf* was an instant success and was able to change the popular image of the wolf. As a result, people in Canada and elsewhere started to oppose wolf hunts.

In 1960 Mowat bought a schooner and, with a small crew, sailed along the coasts of Newfoundland. The voyage and the heroic efforts of the untrained sailors to keep the boat afloat are described in Mowat's funniest book *The Boat Who Wouldn't Float*.

Mowat found the tranquillity of Newfoundland so attractive that he decided to settle there. The scenery was beautiful, the marine wildlife was rich, and the inhabitants were friendly. It seemed to be a perfect place to live. But the harmony proved to be short-lived. In 1967 a pregnant female fin whale entered a lagoon during an extremely high tide and became trapped as the water level lowered. The whale could

---

[3] Mowat, F. (1983): Never Cry Wolf. Bantam Books, Toronto, p. 156.

have left the cove in a month's time, during the next very high tide. But some young people thought otherwise. They thought tormenting the whale and using her for target practice would be fun. They shot at the animal and ripped open her back with a motorboat's propeller. As soon as Mowat was informed about the incident, he tried to stop the cruelty and rescue the whale: he contacted politicians, the authorities, and the media, and wanted to mobilise those local inhabitants who felt sorry for the whale. It is possible that Mowat would have succeeded, had the bullet wounds and propeller cuts not become infected. After 2 weeks the whale died an agonising death. Mowat's relationship with the local community became so spoiled that he left Newfoundland and settled in Port Hope, Ontario. Mowat wrote about the heart-breaking story in his 1972 book *A Whale for the Killing*.

In 1979 Mowat published *And No Birds Sang*, his account of World War Two. The book is considered one of the finest pieces of anti-war literature.

Mowat's next project started with a question about grey whales. Currently grey whales live in the Pacific Ocean but Mowat was curious whether they had once also roamed the Atlantic. Not only did he find out that grey whales had been eradicated from the Atlantic Ocean by whalers, but his investigation led him to shocking data about how countless other animals had become extinct or rare in the North Atlantic and the northeastern parts of America. Mowat started a thorough research on the diminution of life in the region. The result was *Sea of Slaughter*, a comprehensive history of human impact on wildlife during the last 500 years. The book is a tough read. Not because of its length but because of the sad story it reveals. Mowat himself repeatedly stopped writing the book and at times he thought he would not be able to finish it simply because it was nothing more than a horror story. Despite the distressing subject matter, both the book and the film with the same title are important sources for everybody interested in wildlife conservation.

One of the focal points of *Sea of Slaughter* is sealing. As sealing is regarded as a 'tradition' by some Canadians, Mowat received criticism for condemning the practice. While Mowat accepted that seal hunting was an integral part of Inuit culture and was necessary for their survival, he pointed out that the mass slaughter of seal cubs for their pelts is quite a different matter. Mowat was allowed to observe sealing on the spot. What he saw was pure horror, with pups being skinned alive. He concluded that sealing, in its modern form, is 'an almost uncontrolled orgy of destruction conducted by, and for, people who were prepared to commit or to countenance almost any degree of savagery in order to maintain a high rate of profitability.'[4]

In 1985 Mowat was invited to California to deliver lectures and promote *Sea of Slaughter* but was denied entry to the US probably because of his anti-nuclear views and peace activism. Mowat's claim that he had fired his rifle at US strategic bombers flying over his garden decades earlier might also have played a role in his denial. Although his rifle bullets could not possibly reach the bombers, and the story was probably meant as a joke, US authorities did not seem to appreciate Mowat's humour. His satiric account of the affair was published under the title *My Discovery of America*.

---

[4]Mowat, F. (2004): Sea of Slaughter. Stackpole Books, Mechanicsburg, p. 355.

In 1985, Dian Fossey, the primatologist who had been studying and protecting mountain gorillas in Africa for almost two decades, was killed. Mowat looked on Fossey as a kindred spirit. Based primarily on her diary and correspondence, he published Fossey's biography in 1987. *Virunga* (published in the US under the title *Woman in the Mists*) was selected book of the year.

Farley Mowat was one of the leading authors in the conservation movement. Some of his works are among the greatest masterpieces of the green literature. His books about conservation issues have reached millions of readers in dozens of countries. His articles in newspapers and magazines had a considerable mobilising effect. However, his temperament did not allow Mowat to remain 'only' a conservationist writer: he also was an activist. As the president of the Canadian branch of Project Jonah, an anti-whaling organisation, he worked to end whaling. He kept on bombarding authorities with mails and official requests, demanding actions in various conservation matters. He regularly spoke out on green issues and used his political influence in favour of whales, seals, wolves, and other animals. He supported the Sea Shepherd Conservation Society, both morally and financially. Mowat and his wife donated their land to the Nova Scotia Nature Trust and encouraged others to follow their example. Farley Mowat was, and through his writings, continues to be, an uncompromising spokesman for nature. In an interview he said 'I hope we'll achieve a more humane attitude to the other animals who share this planet with us. And if we don't achieve that there's very little likelihood of our own survival.'[5]

Worth Reading

Mowat, F. (1976). *Canada north now: The great betrayal*. Toronto: McClelland & Stewart.
Mowat, F. (1987). *Virunga: The passion of Dian Fossey*. Toronto: McClelland & Stewart.
Mowat, F. (1991). *Rescue the Earth! Conversations with the green crusaders*. Toronto: McClelland & Stewart.
Mowat, F. (1993). *Born naked*. New York: Houghton Mifflin Company.
Mowat, F. (2001). *Walking on the land*. South Royalton: Steerforth Press.
Mowat, F. (2003). *High latitudes: An arctic journey*. South Royalton: Steerforth Press.
Mowat, F. (2005). *People of the deer*. New York: Carroll & Graf Publishers.
Mowat, F. (2009). *Never cry wolf*. Toronto: McClelland & Stewart.
Mowat, F. (2012a). *A whale for the killing*. Vancouver: Douglas & McIntyre.
Mowat, F. (2012b). *Sea of slaughter*. Vancouver: Douglas & McIntyre.

Worth Watching

Sea of Slaughter (1989)
Ten Million Books: An Introduction to Farley Mowat (1981)
Life and Times – The Life and Times of Farley Mowat (1997)
Never Cry Wolf (1983)

---

[5]Farley Mowat on the Seal Hunt. The Globe and Mail, April the 17th, 2008. www.theglobeand-mail.com/opinion/farley-mowat-on-the-seal-hunt/article18448921/

Jane Goodall – A Lifelong Optimist

Around the middle of the twentieth century, despite the scientific acceptance of Darwinian evolution, most people still thought that humans were separate from and superior to nature. Non-human animals were considered to lack emotions and personalities. Even chimpanzees, our closest relatives were regarded as radically different from us. In addition, while chimpanzees were kept in zoos, circuses, and laboratories, practically nothing was known about chimp behaviour in the wild. In 1960, a young and scientifically untrained woman started the first detailed ethological study on free-living chimpanzees. Within a few years her results challenged not only the way we looked at chimpanzees, but also how we perceived our place in nature.

Jane Goodall was born in 1934. When she was one year old, she received a stuffed toy chimpanzee from her father. No one thought how perfectly this present would fit the life of Jane… The young girl preferred to spend time outdoors and was especially interested in animal behaviour. On one occasion she hid for five hours in a henhouse to find out how hens lay eggs. Aged 12, Jane, together with her sister and two friends, started a nature club called Alligator Society: they went on outings and took notes of what they had seen. But the young Jane Goodall was interested not only in natural history, but also in animal advocacy. The Alligator Society established a small museum and raised money from the entrance fees to save old horses from being butchered. Jane also regularly walked a dog who had little opportunity to do physical activities. And she did not tolerate cruelty to animals; on one occasion she confronted boys who were torturing crabs.

Besides being in nature and with animals, Jane Goodall's other pastime was reading. *The Jungle Book*, as well as the stories of Dr. Dolittle and, of course, Tarzan, were among her favourites. She dreamed of living in Africa amidst wild animals.

After leaving school Jane worked as a secretary and later as a waitress. Though everything seemed to be all right with her conventional life, it became evident that she was destined for a very different career. In 1956 Jane accepted the invitation of

© Springer Nature Switzerland AG 2019
L. Erdős, *Green Heroes*, https://doi.org/10.1007/978-3-030-31806-2_23

one of her friends to visit her family's farm in Kenya. She began to save money for the voyage.

Jane Goodall arrived in Africa in 1957. To secure an income, she decided to phone Louis Leakey at the Coryndon Museum in Nairobi to ask for a job. Leakey was a respected palaeoanthropologist who was among the first people to suggest that humans had originated in Africa. Leakey hired Jane as an assistant but it turned out that he had a much better job for her, one the young English girl had always longed for.

Leakey thought that by studying our closest relatives, the great apes, it would be possible to have an idea of the behaviour of early humans. That is why he envisioned long-term studies of chimpanzees in their natural habitat. He only needed to find someone willing to spend a couple of years in the wilderness observing chimps. When he selected Jane Goodall, many thought Leakey had gone mentally ill. How could a fragile English girl live in the forest in the middle of Africa? How would it be possible for someone with no scientific training to conduct in-depth ethological research?

But Leakey's selection criteria were as clear as they were unconventional. The ideal candidate was perseverant, patient, open-minded, loved animals – and was female. Leakey thought women are better observers than men because women tend to pay attention even to the slightest details of behavioural patterns that are usually overlooked by men. In addition, Leakey was convinced that women are more tenacious than men, therefore they are better in carrying out long-term field studies. The fact that Jane Goodall did not have a university degree was no problem for Leakey. On the contrary: to be untrained also meant to be unbiased by scientific theory and prejudice.

In 1960 Jane Goodall started her studies in the Gombe Reserve, located at the eastern shore of Lake Tanganyika. For a short period she was accompanied by her mother, who did the camp chores while Jane was trying to observe the chimpanzees from dawn till sunset. During the first few months it seemed that the naysayers had been right when pointing out that it is impossible to get close to the chimpanzees in the wild. Whenever Jane approached her study subjects, they fled in terror immediately. However, the chimps gradually got used to the presence of the quiet observer, and, after a while, Jane was totally accepted by the chimpanzee community. This offered a unique opportunity to make groundbreaking discoveries.

One of these groundbreaking discoveries came when Jane saw chimpanzees to strip leaves off twigs to 'fish' for termites. The widely accepted belief that only humans are able to make tools was disproved. When Leakey was informed about the observation, he responded with the much-quoted words 'Now we must redefine tool, redefine man, or accept chimpanzees as human.'[1]

Jane Goodall named chimpanzees she met at Gombe. Thanks to numerous articles, books, and films, David Graybeard, Mr. McGregor, Goliath, Flo, Fifi, and all the other chimps became celebrities alongside Jane Goodall. However, at that time

---

[1] Quoted in janegoodall.org.sg/how-to-be-the-next-jane/

it was regarded highly unscientific to name animals. Jane always referred to chimpanzees as 'she' or 'he' instead of the scientifically accepted word 'it.' The fact that Jane kept no emotional distance from her study subjects was unusual. However, no one could doubt that her scientific results turned Jane into the world's foremost expert on chimpanzees. In 1962 she was allowed to start her PhD studies without having a university degree; she earned her PhD in 1965. Jane described chimpanzee social life, family bonds, politics, and hunting. It became clear that chimpanzees are much more similar to us than previously thought. They have emotions and personalities just as we do. Even their gestures are almost identical to ours: chimpanzees kiss, embrace, and pet each other very much like we do.

Over the years Jane's small camp evolved into an international research centre, attracting an increasing number of students and scholars. Today, the Gombe Stream Research Centre is home to the world's longest continuous field study on any animal.

Studying chimpanzees was extremely rewarding, but there have also been some very difficult periods. In 1966 a polio epidemic broke out among the chimpanzees. By vaccinating the chimps, Jane and her husband Hugo van Lawick were able to save many lives, but help came too late for six of the chimps. In 1975 armed guerrillas arrived from Zaire (today the Democratic Republic of the Congo) by crossing Lake Tanganyika. They kidnapped four students and demanded a large sum and the release of some jailed party leaders. Fortunately, subsequent negotiations resulted in the release of the hostages.

Even if Jane Goodall had stopped working in the 1970s she still would belong to the most outstanding characters in the environmental movement, for two reasons. First, she has shown that empathy with the animals does not hinder scientific progress – if anything, it can contribute to a better understanding of the world around us. Second, by showing that there is continuity between humans and chimpanzees she has also shown that there is no chasm between us and the rest of the animal kingdom.

The chimpanzees of Gombe are relatively safe in a protected area. Other chimpanzees, however, are less fortunate. Poaching and habitat loss threatens the survival of chimpanzee populations in many parts of Africa. Chimpanzees kept for entertainment (for example, in circuses or the movie industry) are usually mistreated. But the chimps used for biomedical research have probably the most miserable lives. Locked in tiny, barren cages, they are subjected to electric shocks, social deprivation, and virus infections. For every chimp sold to a laboratory, a zoo, or a private buyer, up to ten others die during the capture or the transit.

For a long time Jane Goodall hesitated to use her fame and influence to help animals. But a 1986 conference on chimpanzees changed everything. Jane was shocked to discover how terrible the situation was, so she decided to devote her life to conservation and animal advocacy. With other scientists she set up the Committee for the Conservation and Care of Chimpanzees with the aim of protecting wild chimps and securing better conditions for captives. Besides chimps, she has also been concerned about all other animals. Since 1986 she has been tireless in lobbying decision-makers, writing articles and books, speaking out on TV, visiting animals in labs, talking to laboratory directors, and giving talks all over the world. She travels 300 days a year to reach a wide audience. Jane appeared in the 2003 documentary *Chattel*, where she argued against cruel tests using animals. Several aspects of animal advocacy are discussed in her books *The Ten Trusts*, *Brutal Kinship*, and *Visions of Caliban*. According to Jane Goodall, the suffering caused by animal experiments could be reduced considerably at once. First, animal tests that are completely useless, without the faintest hope of ever benefitting people, should be banned immediately. Second, for a large number of animal experiments, there are efficient and reliable alternatives that do not use animals. Third, even for tests that are clearly beneficial for people and for which there are as yet no alternatives, the standards could be improved for the captive animals. Sentient and intelligent animals have to spend their whole lives under much worse conditions than the most dangerous human criminals, even though these lab animals are innocent. Millions of people owe laboratory animals their lives, so don't these animals deserve better treatment?

Jane Goodall is a vegetarian and a strong opponent of factory farming. She thinks a real animal lover cannot contribute to the enslavement and mass slaughter of animals. 'Thousands of people who say they "love" animals sit down once or twice a day to enjoy the flesh of creatures who have been utterly deprived of everything that could make their lives worth living and who endured the awful suffering and the terror of the abattoirs – and the journey to get there – before finally leaving their miserable world, only too often after a painful death.'[2] The book *Harvest for Hope* examines how we can help animals, protect the environment, and become healthier by eating responsibly. Jane Goodall heavily criticises fox hunting, trophy hunting, fur trade, keeping cetaceans in aquaria, and using animals for entertainment. She is board member of the Nonhuman Rights Project, which aims to secure fundamental legal rights for nonhuman animals.

In addition to animal advocacy, Jane Goodall is a leading figure in conservation as well. The Jane Goodall Institute carries out conservation-oriented research, the results of which can be used in practical species and habitat preservation. The Institute's approach is community-centred: its projects always include local communities so that protecting nature will also benefit people by improving their livelihoods. This philosophy can guarantee the long-term success of conservation efforts. The Institute supports sustainable agriculture, which provides a reliable source of income and at the same time ensures the protection of wildlife habitat. The Jane Goodall Institute places a great emphasis on increasing the availability of education, clean drinking water, healthcare, and family planning for local inhabitants. Their environmental education programmes inform people about the importance of natural habitats. The Institute operates the Tchimpounga Chimpanzee Rehabilitation Centre, a sanctuary for chimps and mandrills rescued from the illegal bushmeat or pet trade. The Institute's focus is on Africa, but it is also present on other continents, where they work on both local and global issues. In 1991 Jane Goodall started a

---

[2] Goodall, J.; Bekoff, M. (2002): The Ten Trusts: What We Must Do to Care for the Animals We Love. Harper San Francisco, San Francisco, p. xv.

project called Roots and Shoots, specifically designed to inspire the youth to take action for people, animals, and the environment.

The work of Jane Goodall demonstrates perfectly that the protection of animals, nature, and the human environment are complementary actions with clear synergistic relationships. Also, her achievements show that optimism and enthusiasm are contagious and can motivate a large number of people. Of course, Jane Goodall is aware of the tremendous problems we face. Billions of animals and humans are suffering from multiple causes, wildlife is rapidly diminishing, and the state of our environment is deteriorating. But, as Jane put it, 'There is still a lot left that's worth fighting for.'[3] Many of us may sometimes be discouraged by the fact that our everyday decisions can only have a relatively modest impact. But we should keep in mind that all of our small actions add up. If there are millions and millions who do the right thing, we will be able to make the world a better place. That is why the situation is not hopeless. And that is why Jane Goodall continues to urge us to do our best, as citizens, voters, and consumers. 'What you do makes a difference, and you have to decide what kind of difference you want to make,'[4] she holds.

Worth Reading

Goodall, J. (2010a). *In the shadow of man*. Boston: Mariner Books.
Goodall, J. (2010b). *Through a window: My thirty years with the chimpanzees of Gombe*. Boston: Mariner Books.
Goodall, J., & Bekoff, M. (2002). *The ten trusts: What we must do to care for the animals we love*. San Francisco: Harper San Francisco.
Goodall, J., & Berman, P. (1999). *Reason for hope: A spiritual journey*. New York: Warner Books.
Goodall, J., & Hudson, G. (2014). *Seeds of hope: Wisdom and wonder from the world of plants*. New York: Grand Central Publishing.
Goodall, J., McAvoy, G., & Hudson, G. (2005). *Harvest for hope: A guide to mindful eating*. New York: Warner Books.
Goodall, J., Maynard, T., & Hudson, G. (2009). *Hope for animals and their world: How endangered species are being rescued from the brink*. New York: Grand Central Publishing.
Nichols, M., & Goodall, J. (2005). *Brutal kinship*. New York: Aperture.
Peterson, D., & Goodall, J. (2000). *Visions of Caliban: On chimpanzees and people*. Athens: The University of Georgia Press.

Worth Browsing

www.janegoodall.org
www.rootsandshoots.org

---

[3] Jane Goodall in Racing Extinction (2015).

[4] Quoted in team.janegoodall.org/site/PageServer?pagename=ieatmeatless_pledge

Worth Watching

Jane (2017)
Chimps: So Like Us (1990)
People of the Forest (1991)
Jane Goodall: My Life with the Chimpanzees (1995)
Fifi's Boys: A Story of Wild Chimpanzees (1996)
Nature: Jane Goodall's Wild Chimpanzees (1996)
Jane Goodall's Wild Chimpanzees (2002)
Chattel (2003)
Poisoning for Profit (2003)
Jane Goodall's Return to Gombe (2004)
Dead Society (2007)
Jane's Journey (2010)
LoveMEATender (2011)
Surviving Progress (2011)
Me...Jane (2015)
Time to Choose (2015)
Change for Chimps (2016)

No One Loved Gorillas More – The Life and Legacy of Dian Fossey

In 1960 Louis Leakey, the celebrated palaeoanthropologist launched the study of Jane Goodall on free-living chimpanzees because he thought the great apes could provide valuable information on the behaviour of early humans. Goodall's studies turned out to be extremely successful, and Leakey was planning a similar long-term research on mountain gorillas. For this challenging project he needed a person who was no less determined and enthusiastic than Jane Goodall. When he chose Dian Fossey, Leakey made history once again.

Dian Fossey was born in 1932 in California. Her parents got divorced in her childhood, and Dian lived with her mother and stepfather. She hated his stepfather and spent little time with her mother. The young Dian Fossey was shy and lonely. But she loved animals and was able to form especially strong emotional bonds with them. That is why she decided to become a preveterinary medical student. She excelled in most courses, but had problems with physics and chemistry. Finally she decided to devote herself to children with physical or emotional disabilities, so she graduated as an occupational therapist in 1954. She got a job at a children's hospital in Louisville, Kentucky. She was extremely hard-working and dedicated her full attention and creativity to the children. She loved the children she worked with, and they loved her, too. Fossey rented a shabby cottage on a farm outside the city. She spent much of her free time enjoying nature, helping the owners with the daily chores, and looking after the farm animals.

In 1960, one of her friends went on safari to Africa. She asked Dian to join her, but Dian could not afford the travel. However, her desire to watch animals in their unspoiled African environments was so strong that she decided to make the journey within the next few years. She read books about Africa, began to save money, and eventually took out a huge loan. In 1963 her dream to travel to Africa came true. Accompanied by a safari guide, Dian visited iconic places such as Tsavo and the Serengeti. At Olduvai Gorge she even had the opportunity to briefly meet Louis Leakey. The most decisive event of the safari, however, was when Dian Fossey visited the mountain gorillas in the Virunga Mountains. The encounter fascinated Dian

© Springer Nature Switzerland AG 2019
L. Erdős, *Green Heroes*, https://doi.org/10.1007/978-3-030-31806-2_24

so much that she decided to come back to these magnificent beings some time – a plan that seemed highly unlikely.

Once the safari was over, Fossey returned to her job and published articles in the Louisville *Courier Journal* about her African experiences. She even enrolled in a writers' course to improve her writing skills. If and how she would get back to the gorillas remained unclear till 1966.

In March 1966 Leakey delivered a lecture in Louisville. After the event Dian showed Leakey her writings about Africa. Leakey was just looking for someone willing to carry out field studies on mountain gorillas, and he thought the enthusiastic Dian Fossey was a perfect candidate for the job. Leakey told Dian it was necessary to have her appendix removed prior to a long-term stay in a remote African place. Undeterred, Dian had the operation done. Hardly had she arrived home from hospital when she read Leakey's letter explaining that there was no need to have her appendix removed. It was only Leakey's way to test the candidate's determination...

Christmas 1966 found Dian Fossey in Africa, where she spent a few days with Jane Goodall and her husband, Hugo van Lawick. From Kenya Dian had to travel to the Congo, where she was supposed to start her gorilla studies. The American girl had only a very limited knowledge of African geography, culture, local languages, and wildlife. Alan Root, the well-known nature documentary filmmaker offered his help because he thought without some assistance Dian Fossey could not even find the target country, let alone found a research camp.

It was Alan Root who taught the first tracking lesson to the inexperienced Dian Fossey. When they found a gorilla track, Dian excitedly started to follow the trail – in the wrong direction. After a couple of minutes, Alan Root said, 'Dian, if you are ever going to contact gorillas, you must follow their tracks to where they are going rather than backtrack trails to where they've been.'[1]

Fossey and Root established the research camp on the Kabara meadow, where the biologist George Schaller had camped some years earlier. He had spent there one year studying gorillas. Schaller was among the first people who portrayed gorillas as gentle giants rather than monsters.

The beginning was difficult. Dian Fossey had to start a completely new research project under extremely difficult field circumstances, on a high mountain, isolated from civilisation. But she loved nature and could hardly wait to observe her study subjects. Gradually she learned tracking and developed her own method for approaching gorillas. She adopted a submissive posture, imitated the vocalisations of the gorillas, and mimicked eating wild celery. After a while the gorillas accepted her presence.

In July 1967, the political situation, which had already been quite insecure in the Congo, deteriorated further. A rebellion broke out, and the director of the national park ordered Dian to leave her camp. For her own safety, she was escorted to the park headquarters, where she spent roughly two weeks. Nobody knows what exactly

---

[1] Fossey, D. (2000): Gorillas in the Mist. Mariner Books, Boston, pp. 6–7.

happened there. What is certain is that somehow Dian managed to escape from the Congo, even though the border was closed. She arrived in a terrible condition at a hotel on the Ugandan side of the border, which was already full of refugees.

Louis Leakey suggested that Dian should start a study on lowland gorillas or orang-utans. But she wanted to continue her project in the Virungas. So she resumed her work on the Rwandan part of the mountain range. She set up a new research camp on an idyllic natural clearing divided by a brook and surrounded by lush forests, located just a stone's throw away from her previous camp.

Fossey continued her studies with the methods she had developed in the Congo. Within a short time she accumulated an impressive amount of data on the behaviour of mountain gorillas and had important scientific observations about countless aspects of gorilla life. She enjoyed spending time with the gorillas, who also seemed to enjoy her company. Some of the most curious gorillas went to Dian, touched her, examined her equipment, flipped through the pages of *National Geographic Magazine*, or just sit in the proximity. Adult gorillas even allowed Dian to play with the youngsters. Dian Fossey became an honorary member of gorilla society.

Although Dian Fossey's favourites were gorillas, she loved other animals, too. She had a dog, kept a rescued monkey, raised chickens as pets, fed wild birds, and even named giant rats visiting the camp. As the camp and its surroundings were free from poachers, the area served as a refuge for local wildlife, including antelope and buffalo.

Based on her groundbreaking results, Dian Fossey earned her PhD in 1976 at Cambridge University. The camp developed into an internationally renowned research centre, hosting PhD-students, film crews, and other visitors.

However, life in the Virungas for Fossey was more than research: it was also about protecting gorillas. The habitats of the mountain gorillas formed the Volcanoes National Park. Unfortunately, the legal protection existed only on paper. Due to overpopulation, the area of the park was reduced from time to time. Although grazing in the park was prohibited, herders routinely drove their cattle into the park, destroying the vegetation and making the area unsuitable for gorillas. The most serious problem, however, was poaching. Poachers' primary target was usually not the gorillas; they used wire snares to capture antelopes. But the snares were dangerous to gorillas as well; when a gorilla got caught in a snare, only rarely were they able to escape. The noose gradually tightened around the animal's wrist or ankle until, after long suffering, they died of gangrenous infection. Sometimes poachers killed gorillas for their heads and hands, which then could be sold to wealthy westerners as souvenirs. Occasionally, the poachers intended to capture live infant gorillas to sell them to zoos. As adult gorillas defend the infants at all costs, capturing one single young gorilla usually entailed the massacre of a complete family.

Quite naturally, the national park had some park guards. However, they were underpaid, unmotivated, badly equipped, and rarely ventured into the forest. Moreover, the guards typically had good contacts with the poachers, thus the poachers could 'work' within the park unhindered.

It is no wonder that Dian Fossey felt she had to do something about this terrible situation. When she started her studies, there were less than 300 mountain gorillas in the Virunga Mountains, and their number was rapidly decreasing. Dian contacted the authorities and requested that they enforce nature conservation laws. As nothing happened, Dian decided to take gorilla conservation into her own hands. It was clear to her that either she would have to do something to protect the gorillas of the Virungas, or they would become extinct within a short period.

Dian Fossey started what she called active conservation. With her African staff and foreign students working at camp, she chased away cattle, destroyed thousands of traps, rescued and healed trapped animals, captured poachers, confiscated their weapons and equipment, and demolished their temporary shelters in the forest. For equipping her staff with boots and raingears, she had to rely on her own savings, while she could hardly buy food for herself. Though strictly speaking her conservation actions were illegal, even some Rwandan authorities considered them desirable and tacitly approved Fossey's measures. In effect, Fossey was enforcing Rwandan conservation legislation. 'The day the park guards are trained and motivated to do their work, that is the day I will happily stop patrolling. Until then I don't have a choice,'[2] Dian asserted.

Robert Campbell and Ian Redmond were among the most active members of Fossey's anti-poaching patrols. Campbell was a wildlife photographer working for National Geographic, while Redmond arrived at camp as a student interested in gorilla research. Both of them regarded gorilla conservation as extremely important, and so became her most trusted co-workers. Fossey also could rely on her African trackers, who did not hesitate to risk their lives in the fight against poachers. Emmanuel Rwelekana and Alphonse Nemeye were among those who worked the longest time for her.

Dian lived in constant fear for the animals. Some extremely terrible events made her determination and anti-poaching efforts even stronger. One day a buffalo got entangled in a shrub. Poachers found the animal, hacked off his hind legs, and left the poor buffalo alive. On another occasion, the national park hired the most notorious poacher of the region to capture two young gorillas for the Cologne Zoo. Of course, neither the poachers nor the national park staff knew how to keep gorillas. When Dian Fossey was informed about the situation, the babies were already half-dead. Fossey nursed the animals around the clock. The two gorillas, whom she named Coco and Pucker, recovered after a couple of weeks of intensive care. Fossey's plan was to return them to the wild. She contacted the Cologne Zoo, international conservation organisations, Rwandan authorities, and the German embassy. Well-known conservationists such as Peter Scott and Bernhard Grzimek supported Fossey, demanding that the two captives be rewilded, but all the attempts were in vain. The infants were exported to Cologne. The head of the national park received a new Land Rover, some cash, and a travel to Germany. Two gorilla families had to die for that reward. To prevent further kidnaps, Fossey organised an intelligence network to keep her informed about the plans and activities of potential gorilla buyers, traders, and poachers.

---

[2] Quoted in Mowat, F. (1988): Virunga: The Passion of Dian Fossey. Seal Books, Toronto, p. 251.

On New Year's Eve 1977, Digit, Dian's closest friend among the gorillas, was brutally murdered by poachers. His decapitated, handless body was found on January the 2nd in the forest. '...all of Digit's life, since my first meeting with him as a playful little ball of black fluff ten years earlier, passed through my mind. From that moment on, I came to live within an insulated part of myself,'[3] Dian wrote in her diary. Later it turned out that the poachers got a sum equivalent to $20 for Digit's head and hands. Fossey was devastated. Her personality changed and, from that point on, the anti-poaching war became a personal struggle for her. A couple of days after the murder, David Attenborough's team arrived at camp to film the gorillas for the documentary series *Life on Earth*. The national park guards spent their time harassing the BBC crew instead of pursuing the poachers... During the following period, poachers continued to decimate the gorilla population of the Virungas. 'I awakened each morning wondering who would be the next,'[4] Fossey wrote.

Ian Redmond suggested that the massacre should be publicised to raise awareness about the urgency of gorilla conservation. Digit's death was announced on the CBS Evening News. The Digit Fund was created to raise money for anti-poaching patrols (today it operates under the name The Dian Fossey Gorilla Fund International).

Dian Fossey was aware of the importance of gorilla tourism (i.e., organised tours to observe gorillas in the wild) in generating income for Rwanda and informing the public about these majestic animals. Thus, she approved gorilla tourism, provided that it satisfies two basic criteria. First, it should always respect the animals, and second, some of the profit should be used for the protection of the gorillas. Today we call this approach ecotourism.

In 1980 Dian left Africa for a while. She accepted a post at Cornell University. Her course about the great apes was very popular and she was voted the best professor of the year. When, after more than 3 years, she returned to the Virungas, she was happy to see that her gorilla friends remembered her – and embraced her upon her arrival.

In 1983 Fossey's book *Gorillas in the Mist* was published. It was an instant success, has been translated into several languages, and is now widely regarded as a classic. The book is a must for those who are interested in gorillas or African wildlife in general.

While Fossey's scientific achievements and tireless conservation activity brought her international recognition, her single-mindedness also earned her several enemies. She was brutally murdered in her cabin on December the 26th, 1985. Her dead body was discovered the next morning, lying on the floor by the bedside, her skull split by a machete. She was buried alongside her gorilla friends in the gorilla graveyard next to her cabin. The murder is still unsolved.

Though it may seem logical to infer that Dian Fossey was killed by a poacher, this is highly unlikely according to most of her biographers and those who know Rwandan circumstances of that time.

---

[3] Fossey, D. (2000): Gorillas in the Mist. Mariner Books, Boston, p. 206.

[4] Fossey, D. (2000): Gorillas in the Mist. Mariner Books, Boston, p. 209.

Rwandan authorities charged tracker Emmanuel Rwelekana and an American student with the murder. Rwelekana was arrested and committed suicide in prison, though many think he was in fact murdered. The student escaped from Rwanda and was sentenced in absentia to death. But there is almost nobody who believes that the official version was true.

It is almost certain that Dian Fossey had to die because she knew too much about the illegal trade of ivory, gorilla heads and hands, live infant gorillas, bushmeat, and the smuggle of gold. Not long before Fossey's death, her team captured two poachers, who revealed information about those involved in the above business, including poachers, dealers, middlemen, and even some high-ranking officials. Dian Fossey presented a threat to the whole network that made money from illegal activities. Her murder must have been ordered by a powerful member of that network.

Dian Fossey lived a hard life that is unconceivable for most of us. She worked in rugged terrain, at a high altitude, where she had to endure pouring rain and frequent hailstorms. For most of the time she was lonely and her living conditions were Spartan. The political circumstances were unstable, and the forest was full of armed poachers. To make things worse, she was treated with suspicion and hostility by some of her fellow researchers. When she began her studies, the time was running out for the mountain gorillas. Fossey started a heroic struggle; she had a small private army, operated her own intelligence agency, captured poachers and used black magic performances to deter them. Her methods remain controversial, but, according to the most reliable sources (and contrary to some gossips), she never ever hit or tortured poachers. She placed the interests of gorillas above those of her own, and considered conservation more important than data accumulation. Eventually, she gave her life for her gorilla friends. Mountain gorillas are still endangered, but their population is increasing. Dian Fossey was murdered, but she did not lose her fight.

Worth Reading

De la Bédoyére, C. (2011). *Letters from the mist*. London: Palazzo Editions.
Fossey, D. (2000). *Gorillas in the mist*. Boston: Mariner Books.
Gordon, N. (1993). *Murders in the mist. Who killed Dian Fossey?* London: Hodder and Stoughton.
Hayes, H. (1990). *The dark romance of Dian Fossey*. New York: Simon and Schuster.
Mowat, F. (1987). *Virunga: The passion of Dian Fossey*. Toronto: McClelland & Stewart.
Nienaber, G. (2006). *Gorilla dreams: The legacy of Dian Fossey*. New York: iUniverse.
Schaller, G. B. (2010). *The year of the gorilla*. Chicago: The University of Chicago Press.
Shoumatoff, A. (1988). *African madness*. New York: Knopf.

Worth Browsing

gorillafund.org

Worth Watching

The Lost Film of Dian Fossey (2002)
Dian Fossey: Secrets in the mist (2017)
Gorillas revisited with David Attenborough (2005)
Fossey's war (1980)
Search for the Great Apes (1975)
Gorillas in the mist (1988)
Virunga (2014)
The Lost Gorillas of the Virunga (2008)

Biruté Galdikas and the People of the Forest

In Malay orang-utan means 'person of the forest.' Orangutans live only in the swamp rainforests of Sumatra and Borneo. They are largely solitary and mainly arboreal, thus they are even more difficult to observe than chimpanzees and gorillas. Little had been known about the behaviour of orang-utans until the pioneering field-works of Biruté Galdikas, Louis Leakey's third protégé.

Biruté Galdikas was born to Lithuanian refugee parents in Germany in 1946. She was two years old when they immigrated to Canada. She grew up in Toronto. As a child Biruté loved nature and wanted to study orang-utans. 'I was born to be with orangutans,'[1] she says. She was also interested in human evolution. In 1964 her family moved to the US. Galdikas was studying anthropology at the University of California, Los Angeles when he met Louis Leakey after one of his lectures. After an initial hesitation Leakey recognised the young student's strong determination and selected her for the planned studies on orang-utans.

Biruté Galdikas started her orang-utan research in 1971 with her husband Rod Brindamour. Originally they planned to carry out the project in Sumatra, but an Indonesian official suggested that they should opt for Tanjung Puting Reserve, located on the island of Borneo. The reserve was a mostly unexplored territory at the time and was a challenging area for two young westerners.

The couple set up their camp and named it Camp Leakey. They walked through the forest all day long to locate and observe orang-utans. Their hard work and per-severance paid off as they were able to habituate orang-utans to the presence of the human observers. Biruté even made friends with some of the study subjects, includ-ing Ralph, a huge male orang. The scientific value of Biruté's observation is enor-mous. She was the first to record the tool-use of orang-utans. She discovered that adolescent females are quite social and spend some time in groups. She documented

---

[1] Biruté Galdikas in an interview with Tadzio Mac Gregor (Biruté Mary Galdikas: 'If Orangutans Go Extinct, It Will Be Because of Palm Oil'). Huffington Post, August the 12th, 2014. www.huff-post.com/entry/birute-galdikas-if-orangu_b_6055924

© Springer Nature Switzerland AG 2019
L. Erdős, *Green Heroes*, https://doi.org/10.1007/978-3-030-31806-2_25

orang-utan diet, reproduction, social interactions, communication, and movement patterns.

Galdikas earned her PhD in 1978. The scientific community was enthusiastic about her achievements and Galdikas remains a leading authority on orang-utans. But her top priority is conservation. Her career as an activist started as early as 1971 and became progressively more important over the following years.

Hardly had Biruté and her husband settled in Borneo when a baby orang-utan was brought to them. Infant orang-utans were routinely kept as pets in Indonesia, even though the practice was illegal. Biruté could not refuse taking care of them after they had been confiscated. Her aim was to bring up the orphans, teach them how to live in the wild, and release them into the jungle. But serving as a surrogate mother for an orang is a full-time job. Biruté would often follow wild orang-utans in the swamp forest for her studies while rescued baby orangs were clinging to her body. Camp Leakey was taken over by young orang-utans and was developed into a rehabilitation centre. When ready to live on their own, orang-utans are free to leave camp. However, many of them return regularly just for fun or to have human company for a while. Hundreds of orang-utans have been released since 1971. 'I see orangutans that we released 40 years ago still alive and living good lives in the wild,'[2] Galdikas proudly describes the results of her efforts.

In addition to the orang-utan rehabilitation, Biruté has been working on other conservation projects as well. Right from the beginning of her studies, Biruté and her staff, in co-operation with the local authorities, patrolled the reserve thereby contributing to a substantial decrease in poaching and illegal logging activities. The fact that Tanjung Puting Reserve was declared a national park in 1982 was largely due to Galdikas' work.

With the help of some colleagues and supporters, Galdikas created Orangutan Foundation International (OFI) in 1986. The foundation's activity is organised around four focal issues. First, they rescue captive orang-utans, provide them with medical and emotional care, and prepare them for release into the wild. Second, the Foundation works intensively to protect rainforest areas by patrolling existing reserves, lobbying for the designation of new protected areas, restoring degraded forest habitats, and purchasing land to ensure the survival of the rainforest. Third, OFI raises awareness about the plight of the orang-utan by performing educational programmes in schools and for managers and employees of palm oil companies. Last but not least, OFI supports the scientific study of orang-utans, as well as etho-logical and ecological research on proboscis monkeys, gibbons, and red leaf-eating monkeys.

The Foundation is also involved in animal welfare activism. When Galdikas realised that a nearby private zoo was keeping animals in miserable conditions, she urged authorities to close the menagerie. But in the end, the Foundation purchased

---

[2] Biruté Galdikas in an interview with Marsea Nelson, WWF Travel (Q-and-A with Dr. Biruté Mary Galdikas). www.worldwildlife.org/blogs/good-nature-travel/posts/q-and-a-with-dr-birute-mary-galdikas

the zoo. Some of the animals were released, while the circumstances for the others were substantially improved.

The Orangutan Foundation International employs more than 200 local Indonesians. In addition, its facilities belong to the most important tourist attractions of the region, generating considerable revenue for the local population.

Today Biruté Galdikas is an internationally renowned conservationist. She has received a number of prominent awards for her work, including the United Nations Global 500 Award and the PETA Humanitarian Award. She is professor at Simon Fraser University in Canada. But first and foremost, Galdikas is a respected member of Indonesian society, which is a key to the success of her work.

Orang-utans are critically endangered and their population is rapidly declining. Their numbers have been reduced by ca. 50% during the last few decades. The biggest threats to orang-utan survival include the mining of gold and zircon, forest fragmentation due to road constructions, forest fires, and the illegal animal trade. But palm oil is the single most important enemy of orang-utans. 'Orangutans are being driven to extinction by palm oil, it's that simple,'[3] Galdikas says. Palm oil is the cheapest vegetable oil and is therefore a widely used ingredient in food products and cosmetics. Indonesia and Malaysia are the world's largest producers of palm oil. Vast rainforest areas are replaced by oil palm monocultures. Orang-utans and various other species lose their home as one of the most diverse ecosystems on Earth is being destroyed. '… if orangutans go extinct then it will be because of the palm oil industry,'[4] Galdikas warns. As consumers we have a huge responsibility. Orang-utans can be saved if we refuse to buy products that contain palm oil. Checking the ingredients and finding alternative products needs some time, but it is certainly worth it as it can ensure the survival of this wonderful being.

Worth Reading

Galdikas, B. M. F. (1995). *Reflections of Eden: My years with the orangutans of Borneo*. Boston: Little, Brown and Company.

Galdikas, B. M. F., & Briggs, N. (1999). *Orangutan odyssey*. New York: Harry N. Abrams.

MacKinnon, J. (1975). *In search of the red ape*. New York: Ballantine Books.

Montgomery, S. (2009). *Walking with the great apes: Jane Goodall, Dian Fossey, Biruté Galdikas*. White River Junction: Chelsea Green Publishing.

[3] Biruté Galdikas in an interview with Kristina Simona (Dr. Biruté Galdikas: it would be enormous pity if orangutans and rainforests disappeared), Ecopost, June the 16th, 2014. ecopostblog.wordpress.com/2014/06/16/birutegaldikas/

[4] Biruté Galdikas in an interview with Tadzio Mac Gregor (Biruté Mary Galdikas: 'If Orangutans Go Extinct, It Will Be Because of Palm Oil'). Huffington Post, August the 12th, 2014. www.huffpost.com/entry/birute-galdikas-if-orangu_b_6055924

Worth Browsing

orangutan.org

Worth Watching

Search for the Great Apes (1976)
Orangutans: Grasping the Last Branch (1989)
In the Wild: Orang Utans with Julia Roberts (1998)
Born to be Wild (2001)
The Last Trimate (2008)

Jacques-Yves Cousteau – In Awe of the Oceans

Anyone who has had the opportunity to try snorkelling or any other type of diving in tropical or Mediterranean waters will confirm that it is very easy indeed to get captivated by the sea. This is what happened to Jacques-Yves Cousteau, the famous French inventor, explorer, and conservationist.

Jacques-Yves Cousteau was born in 1910 in Saint-André-de-Cubzac, a settlement in the southwest of France. Cousteau's love of the underwater world started when he was ten years old. While horseback riding in a summer camp in the US, he fell off the horse. Although he was not injured, he refused to ride so he was ordered to dive into the nearby lake to remove branches from the water. What was intended as a punishment turned out to be a reward. The young boy took pleasure in the job and could hardly wait to return to the water each day. As a teenager Cousteau was interested in films and writing, which had a lasting effect on his later career.

In 1930 Cousteau enrolled at the naval academy. Because of his heart ailment he would have had no chance to meet the requirements but passed the medical exam due to an unlikely episode: the doctor was about to examine Cousteau when he was called away to an emergency. Cousteau waited patiently but by the time the absent-minded doctor returned he had forgotten where he had left off. The doctor thought he had already examined Cousteau so he ordered him to leave.

However, Cousteau's dream to become a naval pilot vanished when he suffered bad injuries in a car accident. He was advised to take up daily swimming in the Mediterranean Sea to recover faster. The final decision to explore the seas came when, in 1936, a friend lent him a pair of goggles. Cousteau was fascinated by marine life and from this point on turned his full attention to the water world.

In World War Two Cousteau took part in the fight against Nazis. His military duties also involved carrying out underwater experiments. It was during this time that Cousteau, in co-operation with engineer Émile Gagnan, invented Aqua-Lung, the prototype of the modern scuba gear, which started a new era in the study of marine and freshwater environments. After the war Cousteau was active in mine-clearing operations, and he also undertook diving expeditions to explore shipwrecks.

© Springer Nature Switzerland AG 2019
L. Erdős, *Green Heroes*, https://doi.org/10.1007/978-3-030-31806-2_26

In 1950, thanks to the generous support of British politician and businessman Thomas Loel Guinness, Cousteau acquired the ship Calypso. He modified the former minesweeper into an oceanographic research vessel. The Calypso served as Cousteau's headquarters for about 45 years. Also, it was the home of his family. His wife, who loved the sea as much as Cousteau did, spent most of her time aboard the Calypso. Their children learned to swim and dive earlier than to read and write. After they reached school age they were sent to a boarding school, joining their parents aboard the Calypso during longer school breaks.

Cousteau studied the world's oceans, seas, and rivers. Through his films and books, he provided a glimpse into the mysterious underwater world for a huge audience. His 1953 book *The Silent World* quickly became a classic; it has been translated into more than twenty languages and sold millions of copies. The 1956 documentary with the same title won an Oscar for Best Documentary Feature and a Palme d'Or at the Cannes Film Festival. The film was among the first to show colour underwater scenes. Unfortunately, the making of the documentary was extremely destructive: it included the harassment of many animals, the killing of a whale, the massacre of sharks, and the demolition of a coral reef section, together with its inhabitants.

Cousteau received further criticism for the series *The Odyssey of the Cousteau Team*, which represented a strange mix of anthropocentric arrogance and ecological consciousness. In one episode, the team captured two seals (they named them Pepito and Cristobal), to investigate their behaviour in what was nothing more than a pseudo-scientific study. Cousteau himself admitted that he felt the capture had been morally wrong. Nonetheless, the series cemented Cousteau's global fame and his status as a celebrity conservationist.

As he witnessed the rapid deterioration of the marine environment, Cousteau's commitment to conservation deepened, gradually evolving from a peripheral interest and occasional campaigns to his primary and continuous mission.

In 1960 plans were announced to discard radioactive waste into the Mediterranean Sea. Building on his popularity and political influence, Cousteau organised a campaign, which was so successful that the French government had to cancel the plan of using the sea as a dumping ground. In 1971 Cousteau published an influential article in *The New York Times*, warning that the oceans are seriously threatened by overharvesting and pollution. In 1973 he established the Cousteau Society, an organisation devoted to the conservation of marine life.

From the 1970s onwards, Cousteau's films abandoned the terrible impact of his first documentaries. The movie *Voyage to the Edge of the World* as well as the series *Amazon* and *Cousteau's Rediscovery of the World* (also known as *Rediscover the World*) conveyed important conservation messages to the viewers.

In the early 1980s, during the Law of the Sea negotiations, Cousteau advocated the concept of 'zones of responsibility.' The utilisation of these zones would have been controlled by a global ocean authority, possibly contributing to the sustainable management of marine resources. However, Cousteau's idea was rejected, and the concept of 'exclusive economic zone' was accepted instead. Fortunately enough, Cousteau's next campaigns were much more successful.

Cousteau worked hard to convince the International Whaling Commission to end the slaughter of cetaceans, and thus played a central role in achieving the 1986 moratorium on commercial whaling.

In the 1980s, there was an increasing interest in exploiting the resources of Antarctica. Together with numerous other environmentalists, Cousteau started an intensive lobbying while the Cousteau Society launched a petition to save this last wild continent. It was in no small part due to the efforts of Cousteau that France officially sided with the environmentalists, opposing mining and favouring the protection of Antarctica. The next step was to win the support of the USA. The concentrated efforts of Cousteau were backed by Greenpeace, several other organisations, and an increasing number of US politicians, including Al Gore. In 1990 Cousteau organised a highly unusual school trip for six children from six continents – to Antarctica. The children, representing future generations, symbolically took possession of the White Continent. The resulting film *Lilliput in Antarctica* was aired across the US and in several other countries. In addition, Cousteau sent the video to US politicians. At the peak of the campaign, in October 1990, Cousteau and his son Jean-Michel met US President George H. W. Bush, and managed to convince him that Antarctica needs protection, not exploitation. The Madrid Protocol, which prohibits mining on Antarctica, was signed in 1991. It has been one of the greatest successes of the green movement.

Jacques-Yves Cousteau continued to be a spokesperson for nature until his death in 1997. His last book, *The Human, the Orchid, and the Octopus*, was published posthumously in 1998. Cousteau was an adventurer and filmmaker, but became a conservationist. His work is carried on by the Cousteau Society, Cousteau's son Jean-Michel, millions of fans, and all those who strive to protect the Earth's fragile marine and freshwater ecosystems.

The Oceans Today

Till the twentieth century the oceans seemed to have an infinite capacity as both the larder and the trash bin of humanity. But the apparent infinity of the oceans was an illusion: it turned out that marine ecosystems are quite fragile. As ocean advocate Wallace J. Nichols said, '...we've taken too much out of the ocean, we've put too much into the ocean.'[1]

The majority of the world's fisheries are seriously overexploited. Due to the modern technology and the deadliest techniques such as trawlnets and driftnets, fish have nowhere to hide. Billions of animals, from crabs to dolphins, die as bycatch. There is a high risk of losing most if not all coral reefs within a very short time because of global warming, as corals are extremely sensitive to increasing water temperatures. The situation is made worse by the fact that oceans absorb almost one third of the anthropogenic carbon dioxide emission, resulting in acidification. Mangrove forests are being cleared and replaced by harbours, shrimp farms, and tourist facilities.

(continued)

---

[1] Wallace J. Nichols in the 2007 movie 11th Hour.

Huge quantities of agrochemicals, sewage, radioactive waste, oil pollution, and various toxic materials are flowing into the oceans uninterruptedly. Millions of tons of plastic end up in the oceans each year, forming garbage patches and entering the food chain. By 2050, with business as usual, the amount of plastic in the world's oceans will outweigh the amount of fish.

Every one of us can contribute to the protection of the world's oceans and seas. Only purchase certified seafood products with ecolabels such as MSC or Dolphin Safe. Do not buy single-use plastics. Try to favour sustainable tourist facilities and never go to holiday resorts that have been built in environmentally sensitive areas. If you are interested in marine activities such as diving or whale-watching, only take part in programmes that respect the environment and the animals.

Worth Reading

Cousteau, J.-Y., & Dumas, F. (2004). *The silent world*. Washington, DC: National Geographic Society.

Cousteau, J.-M., & Paisner, D. (2010). *My father, the captain: My life with Jacques Cousteau*. Washington, DC: National Geographic Society.

Cousteau, J.-Y., & Richards, M. (1984). *Jacques Cousteau's Amazon journey*. New York: Harry N. Abrams.

Cousteau, J.-Y., & Schiefelbein, S. (2008). *The human, the orchid, and the octopus: Exploring and conserving our natural world*. New York: Bloomsbury.

Worth Browsing

www.cousteau.org
www.oceanfutures.org

Worth Watching

Cousteau's Rediscovery of the World (also known as *Rediscover the World*) (1986–1993)
Amazon (1982–1984)
Voyage to the Edge of the World (1976)
Lilliput in Antarctica (1990)
Jacques-Yves Cousteau: My First 85 Years (1995)
My Father, the Captain: Jacques-Yves Cousteau (2011)
The Silent World (1956)

Mike Pandey – Using the Power of Motion Picture

Ideally, nature documentaries should have a strong conservation message. If the message is delivered in a way that resonates with the emotions of the audience, the film may be able to trigger dramatic changes. Mike Pandey, India's best-known filmmaker has repeatedly shown how such achievements are possible.

Mike Pandey was born in Nairobi, Kenya to Indian parents. Their house being located next to Nairobi National Park, the family's neighbours were elephants and lions. Surrounded by rich wildlife, the young boy was fascinated by nature and spent much of his free time in the park. When he was seven he received a camera as a birthday present from his uncle. Taking photographs became his passion immediately.

Together with his brother, Mike spent some years in the United States where he worked as an intern in Hollywood. He settled in India in the early 1970s, and joined the film industry. He was responsible for the special effects in some historical and romance movies. However, his love of nature made him move towards wildlife documentaries. In 1973 he founded Riverbank Studios, which started to create nature films.

Rather than making usual natural history films, Pandey wants to use the power of motion picture to encourage viewers to take action. He regards film as a medium that is able to change people's attitudes towards nature. He wants the audience to feel uncomfortable and worried about how humans threaten the natural world.

Mike Pandey is in particular awe of elephants. His 1994 film *The Last Migration*, which won the prestigious Wildscreen Panda Award, explores a conflict between elephants and humans. As a result of rapidly increasing human population in India, natural areas are under pressure. Forests are being cleared to make land for agriculture, causing the elephants' habitat to shrink. In search for food or new territories, elephants usually destroy croplands and enter villages, occasionally even killing people. As elephants are protected, a possible solution is to capture them, after which they can be relocated to reserves. Alternatively, the captured animals may be trained and subsequently used in forestry or conservation management. The film is objective, showing sympathy for the animals while also understanding the situation

L. Erdős, *Green Heroes*, https://doi.org/10.1007/978-3-030-31806-2_27

of the villagers. Pandey makes clear, however, that the source of the conflict is human encroachment into nature.

Whale shark is the largest fish species. Feeding entirely on plankton and other small organisms, this gigantic animal can be up to 18 m long and can weigh as much as 15 tons. Pandey was nine when he first saw whale sharks, during a voyage from Kenya to India. The next time he met the species was some decades later. The whale shark was lying on the shore, still alive when fishermen were hacking it. When Pandey saw the fading light in the shark's eyes, he decided to do something about the situation. In *Shores of Silence*, one of Pandey's most moving films, the crew documented how these magnificent creatures were hunted. The meat and the fin were exported. As usual, local fishermen received almost nothing for their work, while most of the profit went to the merchants. Pandey suggested that the sharks and the other wildlife of the region could attract a large number of tourists and divers, securing the survival of this unique species and a decent income for local residents. For *Shores of Silence*, Pandey received numerous awards, including his second Wildscreen Panda Award. More importantly, the film brought about legislative changes and was directly responsible for the whale shark being declared a protected species in India. Also, whale shark was included in the Washington Convention (also called CITES), restricting the international trade of this species.

Pandey returned to the topic of elephants in his 2004 film *The Vanishing Giants*, which earned him his third Wildscreen Panda Award. This time he focused on the capture and documented how inhumanely the elephants were treated. Capture programs were intended to protect the elephants by relocating them. However, in practice, the outdated methods, coupled with human cruelty and negligence, caused immense suffering and even the deaths of several animals. The film shows a heartbreaking story and applies powerful and shocking images. After the film was broadcast on television, a national outrage followed and a petition was started, which eventually led to legislative changes. The barbaric methods of elephant capture were banned and strict standards were introduced to ensure that elephants are treated humanely. Also, non-governmental organisations were granted the right to observe capture operations.

In 2006 Pandey turned to a rather unpopular group of animals. *The Vanishing Vultures* traces how vulture populations collapsed due to the veterinary drug diclofenac. If cattle or other animals die shortly after being treated with diclofenac, vultures get poisoned. Stressing the ecological importance of vultures, Pandey urged a ban on diclofenac. The film is believed to have played a major role in India's decision to phase out diclofenac.

Mike Pandey was the producer and cameraman of *Kalpavriksha*, a film that shows how medicinal plants and the traditional wisdom on their uses is vanishing. Pandey has produced *Earth Matters*, a popular television series to promote environmental causes. Following the success of the series, children formed thousands of nature clubs across India. Pandey set up the Earth Matters Foundation, the goal of which is to inform people about conservation issues and inspire them to act.

Pandey has recently launched campaigns to promote the conservation of tigers and horseshoe crabs. His aim is to explain people why these animals and their habitats are important to protect. Also, Mike Pandey is a strong proponent of forest conservation and action against climate change.

> **The Plight of the Tigers**
> Once tigers roamed wide areas in Asia, but today the species is endangered and numbers less than 4000 individuals in the wild. Despite the efforts of conservationists, population numbers are declining. The causes for the decline are multiple. Habitat destruction is one of the primary threats. It is estimated that over 90% of the original tiger habitat has been lost. Natural habitats are shrinking due to logging, agriculture, and climate change. Habitat fragmentation is a related problem. Fragmentation means that only small and isolated habitat patches remain, which are not large enough to support viable tiger populations. With their habitats disappearing, tigers are forced to human-dominated areas, where they come in conflict with local residents, who often kill them. Poaching is an immediate threat to tigers. This magnificent animal is hunted for its skin. Also, most parts of the tiger are attributed magical healing power and are used in medicine. For example, many believe that the tiger's penis enhances male virility. Recently there have been concentrated efforts to increase tiger populations. As Mike Pandey has pointed out, it would be a shame if humans could not save the tiger from extinction.

Mike Pandey filmed an incredible variety of animals, from beetles to turtles to snow leopards. The primary aim of Pandey's films has always been to mobilise people and bring about legislative changes. Pandey has received some 300 awards for his work and was named Hero of the Environment by *Time* magazine. But he is most proud of the difference his films have made. 'The only real reward I seek for the work I do is to see wildlife and wild habitats a little more secure,'[1] Pandey says.

Worth Browsing

www.earthmattersfoundation.org
www.riverbankstudios.com

Worth Watching

Shores of Silence: Whale Sharks in India (2000).
The Last Migration (1994)
The Vanishing Giants (2004)

---

[1] Mike Pandey in Sanctuary Asia, Vol. 29, No. 6, December 2009. www.sanctuaryasia.com/people/earth-heroes/9344-mike-pandey.html

David Attenborough – The Grand Old Man of Natural History Filmmaking

David Attenborough's name has been associated with natural history films for more than six decades, and his work has inspired generations of filmmakers, nature lovers, and scientists all over the world. Still active in his nineties, he is widely considered one of the leading conservationists of our time.

David Attenborough was born in 1926. As a young boy he was fascinated by nature and spent much time observing animals and collecting fossils, stones, and other natural objects. He graduated from the University of Cambridge, where he had studied natural sciences. In 1949 he started working as an editor for a publishing company. However, he soon realised that he wanted to do something different. So he applied for a job at BBC Radio, but was rejected. Fortunately enough, he was offered a position in BBC's television branch. At that time, television was in its infancy. Like most people in the UK, Attenborough himself did not have a TV set; in fact he had watched only one television programme previously.

Attenborough participated in the production of various programmes, some of which were connected to natural sciences. His first great success, however, came with *Zoo Quest*. The series showed adventurous expeditions that captured live animals for London Zoo (collecting wild animals was accepted then but has long been abandoned by modern zoos). With *Zoo Quest* and its sequels Attenborough was able to visit remote places that were virtually unknown for westerners, and the team recorded species and natural phenomena that had never been filmed before.

In 1965 Attenborough accepted a job in the management of BBC. He oversaw the transition to colour broadcasting and turned BBC 2 into a popular channel. After eight years he resigned from his post to return to wildlife films.

Attenborough's first landmark series was made in the second half of the 1970s. *Life on Earth* had the ambitious goal of tracing the history of life on our planet. Filmed over three years in 39 countries, the series succeeded in providing a comprehensive yet understandable and enjoyable summary of the variety of life from an evolutionary perspective. *Life on Earth* revolutionised natural history filmmaking and is estimated to have been watched by half a billion viewers. The series also introduced Attenborough-style narration, which has remained extremely successful

© Springer Nature Switzerland AG 2019
L. Erdős, *Green Heroes*, https://doi.org/10.1007/978-3-030-31806-2_28

ever since: Attenborough appears on the scene from time to time for short periods, and speaks a few words about what we will see. His clothes and gestures lend the film additional authenticity and enable us to become part of the story – almost as if the viewers themselves were in the wilderness with Attenborough. In the twelfth episode, Attenborough encountered mountain gorillas belonging to one of Dian Fossey's study groups in the Virunga Mountains. The sequence is regarded as one of the greatest moments in the history of television.

The 1984 series *The Living Planet*, Attenborough's next major project, focused on the main habitats of the Earth, from the jungles to the polar ice caps, from the depths of the oceans to the highest mountains. The trilogy was completed by *The Trials of Life*, which examined the behaviour of animals. The three series are undoubtedly among the greatest classics of wildlife films as they brought nature closer to people – and probably also people closer to nature.

By the 1980s Attenborough had become a household name worldwide. Each week millions watched him appear in the most unusual places: be it a mountaintop, a tree canopy, a coral reef, a hot air balloon, a sand dune, or a penguin colony. Despite his earlier successes, the 1995 series *The Private Life of Plants* seemed a somewhat risky enterprise, as plants are thought to be less active and, consequently, less exciting than animals. However, Attenborough and his team managed to make a breathtaking series about botany, unparalleled by any other documentary about the world of plants.

The Life of Birds and *The Life of Mammals* were about two groups that had been extensively covered by countless earlier films yet Attenborough was again able to enchant viewers with two perfect series. The 2005 production *Life in the Undergrowth* introduced people to the fascinating but mostly hidden world of invertebrates, including the most bizarre creatures on Earth. Invertebrates are usually neglected, but Attenborough reminds us of the essential role these small beings play in the health and functioning of the ecosystems. Moreover, the series makes it clear that, similar as the individual invertebrates within a species may seem, they have personalities. 'They all have personalities, there's no doubt about it,'[1] Attenborough insists. Amphibians and reptiles are regarded by many as either repulsive or dangerous. *Life in Cold Blood* aims to evoke sympathy for frogs, snakes, and their relatives by showing how interesting and, indeed, important they are.

Life in Cold Blood closed the Life series, which had begun some three decades earlier with *Life on Earth*. But Attenborough has continued with some additional remarkable natural history films, including *Kingdom of Plants* and *Attenborough's Paradise Birds*.

The enduring popularity of Attenborough's documentaries may be due to several factors. The carefully edited scripts are able to convey scientific value in an entertaining manner. In addition, the crew has always placed a great emphasis on using

---

[1] www.bbc.co.uk/pressoffice/pressreleases/stories/2005/10_october/20/life_secrets.shtml

the latest filming techniques such as slow motion and time-laps photography, star-light or infrared cameras, and underwater filming. Another key factor is the person-ality of Attenborough, who perfectly combines the characteristics of the trustworthy expert and the enthusiastic amateur.

Most of Attenborough's documentaries contain important conservationist mes-sages. But some of his films are completely devoted to the environmental cause. *State of the Planet* is probably one of the most important among them. In this three-part miniseries Attenborough identifies the main causes of the current species extinction crisis and summarises the possible solutions. The films *Are We Changing Planet Earth?* and *Can We Save Planet Earth?* examine global climate change and urge immediate action to reduce fossil fuel consumption. 'If we miss our chance, the future may be grim indeed for life on planet Earth,'[2] Attenborough warns. The 2019 documentary *Climate Change: The Facts*, presented by Attenborough, covers the same topic in an even more powerful way, ringing the alarm bells once again. *The Death of the Oceans?* investigates the dangers threatening marine life and stresses the importance of protecting oceans.

Attenborough presented the 2009 documentary *How Many People Can Live on Planet Earth?* He was a perfect choice for the programme, as he considers over-population one of the most serious threats to the natural world – and ourselves as well. There is no doubt that human overpopulation is a key component of virtually every environmental problem.

A Few Words About an Overcrowded Planet

Nine thousand years ago the total human population of the world was about ten million. It took thousands of years till 1800 for our population to reach one billion. Only 120 years later it reached two billion and it took no more than 40 years to add another billion. Today we are nearing eight billion. Each sec-ond our population is increasing by two people.

Human population growth has long been in the focus of scientific interest. As early as the eighteenth century Thomas Malthus warned that food produc-tion would not be able to keep pace with the increase in population. There was a period when human population growth surpassed exponential growth. A 1960 calculation conducted by Heinz von Foerster, Patricia M. Mora, and Lawrence W. Amiot yielded an interesting result.[3] Published in *Science*, their study concluded that, provided that the trend continued, human population

(continued)

---

[2] David Attenborough in the 2006 documentary Can We Save Planet Earth?

[3] von Foerster, H.; Mora, P. M.; Amiot, L. W. (1960): Doomsday: Friday, 13 November, A.D. 2026. Science 132, pp. 1291–1295.

would approach infinity on Friday, November the 13th, 2026. That would be a very unlucky Friday the 13th indeed...

Overpopulation was in the centre of the research of notable ecologists such as Paul Ehrlich, who published the famous book *The Population Bomb*, and Garrett Hardin, whose article entitled *The Tragedy of the Commons* has remained a much-cited work ever since its publication in 1968.

In developed nations human population seems to be stabilised or even declining, while less developed countries face a population explosion. Generally the birth rates are the highest in regions where food security, water security, and the quality of life are the most problematic. Policies and taxes that encourage smaller families, the education and empowerment of women, the availability of contraceptives, and family planning guidance may be useful in reducing the population boom. We should consider the words of David Attenborough: 'Instead of controlling the environment for the benefit of the population, perhaps it's time we control the population to allow the survival of the environment.'[4]

By the mid-2010s Attenborough was usually referred to as the most travelled person in human history. But there still were a few places that he had not yet visited, including the White House in Washington. This, however, changed in 2015, when US President Barack Obama invited David Attenborough for an interview to discuss global problems such as climate change, overpopulation, alienation from nature, and coral die-off. The extraordinary meeting was the subject of the television programme *David Attenborough Meets President Obama*.

Besides helping millions to appreciate the beauty of nature and motivating them to contribute to its protection, Attenborough has also done his bit in environmental activism. He has repeatedly participated in fundraising projects for endangered animals and has supported conservation organisations' programmes, including the WWF's efforts to protect tropical rainforests and BirdLife International's campaign to save albatrosses. Attenborough also backed Cruelty Free International, urging an end to the use of non-human primates in cruel and unnecessary experiments. In 2013 Attenborough joined the pro-badger campaign launched by Queen guitarist Brian May. Badgers can transmit bovine tuberculosis and the badger cull is intended to fight the disease. However, scientific research indicates that vaccination is a more efficient, more humane and less expensive method than the slaughter of thousands of badgers. At the 2018 UN climate summit in Poland, representing the world's people, Attenborough addressed the delegates and asked them to take action in order to save the civilisation and the natural world.

---

[4]David Attenborough in the 2002 documentary The Life Of Mammals, episode 10 (Food For Thought).

Attenborough has been voted the most trusted and the most popular personality in the UK. He holds 32 honorary degrees from British universities alone. About a dozen species have been named in his honour. He has countless admirers all over the world, including former US President Barack Obama, former UK Prime Minister David Cameron, musician Sting, and Prince William. In short, Attenborough is one of the most respected and most beloved individuals on the globe. Clearly, we should follow his lead in doing our best to protect the natural world. Because, as Attenborough put it, 'The future of life on Earth depends on our ability to take action.'[5]

Worth Reading

Attenborough, D. (1979). *Life on Earth*. London: Collins-BBC.
Attenborough, D. (1984). *The living planet: A portrait of the Earth*. London: Collins-BBC.
Attenborough, D. (1992). *The trials of life*. London: Collins-BBC.
Attenborough, D. (1995). *The private life of plants*. London: BBC Books.
Attenborough, D. (1998). *The life of birds*. London: BBC Books.
Attenborough, D. (2002a). *Life on air*. London: BBC Books.
Attenborough, D. (2002b). *The life of mammals*. London: BBC Books.
Attenborough, D. (2005). *Life in the undergrowth*. London: BBC Books.
Attenborough, D. (2008). *Life in cold blood*. London: BBC Books.
Attenborough, D., & Fuller, E. (2012). *Drawn from paradise*. London: HarperCollins.

Worth Watching

Life on Earth (1979)
The Living Planet (1984)
The Trials of Life (1990)
The Private Life of Plants (1995)
The Life of Birds (1998)
State of the Planet (2000)
Life on Air (2002)
The Life of Mammals (2002)
Life in the Undergrowth (2005)
Are We Changing Planet Earth? (2006)
Can We Save Planet Earth? (2006)
Life in Cold Blood (2008)
How Many People Can Live on Planet Earth? (2009)
The Death of the Oceans? (2010)
Attenborough's Ark (2012)
Attenborough: 60 Years in the Wild (2012)
Kingdom of Plants (2012)
Attenborough's Paradise Birds (2015).
David Attenborough Meets President Obama (2015)
Attenborough at 90: Behind the Lens (2016)
Climate Change: The Facts (2019)

---

[5] David Attenborough in the 2000 documentary State of the Planet, episode 3 (The Future of Life).

Paul Watson – The Daredevil of Conservation

'You must be prepared to risk all, including your life and liberty, to uphold the sacred integrity of the Earth,'[1] Paul Watson maintains. And he walks his talk. As a radical conservationist he thinks the forces destroying the biosphere are so aggressive that immediate confrontational action must be taken to halt them. He is admired by conservationists and animal lovers all over the world. His friends and supporters range from Pamela Anderson to the Dalai Lama and include, among others, Sean Connery, Christian Bale, Martin Sheen, Pierce Brosnan, Brigitte Bardot, and Bob Barker. According to writer David B. Morris, 'Paul Watson seems an almost impossible figure: a cross between Saint Francis and Rambo.'[2] Canadian author and conservationist Farley Mowat characterised Paul Watson as 'the most effective defender of wildlife.'[3]

But Paul Watson's popularity is by no means universal. Some regard him as a terrorist. Whalers and sealers hate him. No one knows how many times Watson has been jailed, assaulted, and threatened.

So, who is Paul Watson? 'I am a conservationist and that is my business: getting in trouble. I'm here to say things people do not want to hear and to do things people do not want to see,'[4] he says.

Paul Watson was born in 1950 in Toronto. When he was five the family moved to the east coast and Paul grew up having daily contact with the ocean. At the age of nine he spent much time swimming with a friendly beaver in a nearby pond. When the beaver was killed by trappers, the young boy was terribly sad – and terribly furious. With his brothers and sisters he started to comb the area for traps, released the animals, and destroyed the traps. From the age of 17 he served as a seaman on

[1] Watson, P. (2012): Earthforce! An Earth Warrior's Guide to Strategy, 2nd edition, p. 28.

[2] Morris, D. B. (1995): Earth Warrior: Overboard with Paul Watson and the Sea Shepherd Conservation Society. Fulcrum Publishing, Golden, p. 13.

[3] Quoted in planetenreiter.de/whale-wars-gedanken-zu-einer-bemerkenswerten-doku-serie/

[4] Paul Watson in the 2008 film Pirate for the Sea.

© Springer Nature Switzerland AG 2019
L. Erdős, *Green Heroes*, https://doi.org/10.1007/978-3-030-31806-2_29

various ships. When he co-founded Greenpeace he was one of the youngest and one of the most active members of the organisation.

In 1973, accompanied by fellow Greenpeace activist David Garrick, Watson joined the protest of Native Americans in Wounded Knee. Watson crawled into the settlement that was cordoned off by the authorities and served as a medic during the conflict.

Paul Watson was a central figure in Greenpeace's anti-whaling activity. He was one of the activists who manoeuvred inflatable rubber boats between the whalers and the whales, while harpoons were fired over their heads. One of the defining moments of Watson's life happened when he looked into the eye of a dying whale. Watson vowed he would dedicate his life to protecting ocean wildlife.

Watson was instrumental in launching the anti-sealing campaign of Greenpeace. In 1976 Paul Watson and Robert Hunter stood motionless on the ice, directly in front of an icebreaker sealing ship, to block its movement. A year later Watson confronted sealers again. When a sealer raised his club to kill a seal pup, Watson grabbed the club and throw it into the water. Then, to protest the slaughter of the defenceless baby seals, he handcuffed himself to a pile of pelts tied to a sealing ship. Sealers aboard the ship dragged Watson across the ice, lifted him into the air and then plunged him into the icy water. Sealers seemed to enjoy the incident. Watson almost died of hypothermia.

In 1977 Paul Watson left Greenpeace and set up a new organisation, which would evolve into Sea Shepherd Conservation Society. In the following year, with the support of Cleveland Amory, Watson was able to purchase a ship. The first mission of the newly acquired vessel's crew was to protect baby seals in Canada. Watson and his friends sprayed the newborn seals with a dye that is harmless to the seals but makes their pelts commercially worthless.

The next mission was to find and disable the notorious pirate whaling ship Sierra, which was harpooning whales illegally, totally disregarding international regulations. Its crew harpooned whales indiscriminately, not sparing endangered species, mothers or calves. They kept only the most valuable parts of the whales and dumped the rest into the water. As the outlaw ship could be virtually anywhere on the world's oceans, finding it was like looking for a needle in a haystack. But in 1979 Watson spotted the Sierra off the coast of Portugal and rammed it. 'That ship had already killed 25,000 whales. I wanted to make sure it wouldn't kill any more,'[5] Watson said. The conflict received worldwide publicity and the position of pirate whaling became untenable. At long last the bloody story of the Sierra ended.

In 1981 Watson filmed illegal whaling operations of the Soviet Union and, chased by the Soviet Navy, managed to return to US waters with the evidence. In 1983, using his vessel Sea Shepherd II, Watson blockaded the harbour mouth of St.

---

[5] Quoted in Ryan, M. (1979): In a Dramatic Duel at Sea, a Young Conservationist Rams a Ship to Save the Whales. People, August the 20th, 1979. people.com/archive/in-a-dramatic-duel-at-sea-a-young-conservationist-rams-a-ship-to-save-the-whales-vol-12-no-8/

Whaling, Sealing, and Shark Finning

Whaling has been practiced for millennia, but for a very long time it was limited to coastal waters. As the ships got larger, whaling became more intensive. Steamships made it possible to effectively pursue and kill whales everywhere. Other technological innovations such as the explosive harpoon head were bad news for whales. In the twentieth century several cetaceans were driven to the brink of extinction.

Whales were primarily hunted for their meat, whale oil, and baleen. Whale meat was (and in some countries still is) used for human consumption, the production of animal food, or as fertiliser. Whale oil was utilised for lighting and lubrication, and later to manufacture margarine and various cosmetics. Baleen was used to make fishing rods, umbrella ribs, and a number of other products.

When it became clear that unregulated whaling would deplete whale stocks within a short period, the International Whaling Commission (IWC) was set up in 1946 to secure the long-term utilisation of whales. Its regulations were largely inefficient as the quotas it imposed were too high. However, encouraged by conservation organisations such as the WWF and Greenpeace, several nonwhaling nations (many of them landlocked) joined IWC, which, at long last, banned commercial whaling. Despite the ban, Iceland, Japan, Norway, and South Korea continue whaling. Japan hunted whales for 'scientific purposes,' although it was clear to everyone that the reason was not science. In 2014 the International Court of Justice ruled that Japan's whaling did not qualify as scientific research and that Japan violated the prohibition on commercial whaling in the Southern Ocean Whale Sanctuary. Also, the Court of Justice ordered Japan to stop whaling. In 2018, after their proposal to allow commercial whaling was rejected once again in the International Whaling Commission, Japan announced that it would withdraw from the Commission and resume commercial whaling.

While some countries defend whaling, the overwhelming majority opposes it. As it may take up to 45 min to kill a whale, and the animals die in agony, killing whales is morally unacceptable for most people.

Sealing is surely among the most barbaric practices ever carried out by the human species. While killing seals for their meat and pelts was necessary for the survival of some native communities, the current industrial-scale killing of seal pups has nothing to do with subsistence. Each year hundreds of thousands of helpless newborn seals are massacred, most of them in Canada. Video footage shows unequivocally that many baby seals are skinned alive. Seal pelt is used in the fashion industry to make profit for unscrupulous designers and retailers who provide thoughtless people with luxury though unethical products.

Sharks are less popular but ecologically no less important than whales and seals. Though sharks normally do not present a threat to humans, some species are undeniably dangerous: each year up to ten people die because of sharks. And each year up to 100 million sharks die because of humans – the figure may be as high as 300 million, if illegal fishing is taken into account.

(continued)

Sharks are fished for their meat and culled during 'shark control programmes' that aim to reduce their populations. Also, many sharks are killed unintentionally, as bycatch. But one of the most imminent threats to several shark species is finning. The fins of the sharks are usually cut when the animals are still alive. The sharks are then discarded into the sea.

If you are disgusted with whaling, sealing, and shark finning, never buy products that support these dirty businesses.

John's, Canada, announcing that he would ram any sealing ship trying to leave the port. The action saved the lives of 76,000 seals, as the sealing fleet was not able to depart for two weeks.

In 1984 Canadian authorities seized the ship of the Sea Shepherd Conservation Society. This made Paul Watson focus on the protection of wolves, which were being hunted mercilessly in British Columbia, Canada. His campaign revealed the real causes behind the wolf hunt. By the early 1980s intensive trophy hunting resulted in a significant decrease of game populations. Rather than reducing recreational hunting, British Columbia's minister of the environment masterminded a plan to blame the wolves. The project was called 'game management' or 'conservation,' but its real purpose was to eradicate the wolves of the region. Watson's campaign received heavy media coverage and was instrumental in the resignation of the minister of the environment.

In 1986, after the Society's ship was returned, Watson and his team interfered with the Faroese massacre of pilot whales. The hunt as well as Sea Shepherd's conflict with the local authorities was filmed by a BBC crew, resulting in the documentary *Black Harvest*. In the very same year, two Sea Shepherd activists, Rod Coronado and David Howitt sank two Icelandic whaling ships and destroyed a whale-processing facility, thereby crippling Icelandic commercial whaling activities for a very long time.

During the following years, the Society's main activity was the fight against driftnetting, an extremely destructive fishing technique that kills a huge amount of nontarget species, including dolphins, whales, seals, seabirds, and turtles. Also called curtains of death, driftnets may be dozens of kilometres long, hanging from the surface and extending a couple of metres into the water. They are almost invisible under the water and kill everything that becomes entangled in the net. Lost or discarded nets ('ghost nets') continue to drift around in the ocean, destroying marine wildlife for decades, maybe even centuries. By removing driftnets, Sea Shepherd activists save countless sea animals.

In 1994, while trying to confront Norwegian whaling operations, She Shepherd's ship was attacked by the Norwegian Navy. Though the naval vessel rammed Watson's ship, fired at the activists, and deployed depth charges, the conservationists managed to escape unharmed.

In 1997 Paul Watson was invited to a meeting of the International Whaling Commission. As soon as he entered the room, the Icelandic, Norwegian, Japanese, and Caribbean delegations walked out in protest.

The members of the Sea Shepherd Conservation Society regularly take part in oil spill clean ups. In 2001, after a tanker had run aground in the Galapagos Islands, Sea Shepherd volunteers assisted the staff of the Galapagos National Park in the wildlife rescue.

In 2003 Sea Shepherd volunteers documented the notorious dolphin slaughter in Taiji, Japan, where dolphins are herded into a cove, and are slaughtered for their meat or captured for sale to zoos and aquaria. Sea Shepherd activists Allison Lance and Alex Cornelissen freed fifteen dolphins, for which both activists were arrested and spent three weeks in jail. They were released after Sea Shepherd paid a fine of $8000. Thanks to the efforts of Sea Shepherd and other conservation and animal advocacy groups, the Taiji dolphin hunt continues to receive international media attention. The slaughter is the topic of the Oscar-winning documentary *The Cove*.

In the early 2000s, the fight against Japanese whaling around Antarctica became Sea Shepherd's principal focus. The fleet of Sea Shepherd was chasing and harass-ing the Japanese ships, keeping them from whaling. The conservationists rammed the whalers, deployed propeller foulers, and throw stink bombs on the decks of the whaling ships. The stink bombs contained butyric acid, a harmless and biodegrad-able substance with an extremely disgusting smell. Sea Shepherd's fight to stop whaling is featured in the documentary *At the Edge of the World* and the Animal Planet reality series *Whale Wars*.

Paul Watson has several inspirational sources, but Edward Abbey and Dave Foreman rank among those who have had the greatest influence on Watson's strat-egy. Edward Abbey was a writer and conservationist whose best-known work is *The Monkey Wrench Gang*, an excellent and now classic novel about eco-saboteurs. Dave Foreman is a radical conservationist, co-founder of the organisation Earth First! and editor of the book *Ecodefense: A Field Guide to Monkeywrenching*.

Paul Watson is usually criticised because his tactics are considered violent. But the reality is that the Sea Shepherd Conservation Society has never caused an injury to anybody. Their approach is 'forceful non-violence,' meaning that they always take the greatest care to avoid injury to any living being. Inflicting damage on inani-mate objects, machinery, and equipment does not qualify as violence as long as no one's life or physical integrity is threatened. What is violent is the slaughter of ani-mals for profit and the destruction of the environment in which we live. Forceful non-violence, according to Paul Watson, is the most efficient way to protect the biosphere, and perhaps the only way to enforce national and international environ-mental regulations that no country wants to enforce.

Critics like to point out that killing whales is not different from killing pigs or cattle. This criticism misses the point. Firstly, several whale species (and many other sea creatures) are endangered, while pigs and cattle are not. Secondly, killing a whale is usually a long process, causing immense suffering for the animals, which is clearly unacceptable. Thirdly, and most importantly, there is some undeniable similarity between killing whales and raising (and butchering) livestock: both are cruel to the animals and disastrous from an ecological perspective. But from this fact it does not follow that both practices are morally acceptable. Quite on the contrary: both should be challenged. That is why Paul Watson is a vegan, as are all the ships of the Sea Shepherd Conservation Society.

Paul Watson has been a key figure in the conservation movement for the last 50 years. He has saved the lives of millions of living beings and his work has contributed to significant improvements in environmental legislation. The Sea Shepherd Conservation Society is one of the most effective conservation organisations. They operate a small navy that is dedicated to the protection of the oceans. And if you call them pirates, Paul Watson will not be annoyed. 'We're the good pirates,'[6] he says.

Worth Reading

Ligniti, E. (Ed.). (2018). *Sea Shepherd 40 years: The official book*. Milano: Skira.
Morris, D. B. (1995). *Earth warrior: Overboard with Paul Watson and the Sea Shepherd Conservation Society*. Golden: Fulcrum Publishing.
Watson, P. (1993). *Earthforce! An earth warrior's guide to strategy*. La Canada: Chaco Press.
Watson, P. (2003). *Seal wars: Twenty-five years on the front lines with the harp seals*. Buffalo: Firefly Books.
Watson, P., & Rogers, W. (1982). *Sea Shepherd: My fight for whales and seals*. New York: W. W. Norton and Company.

Worth Browsing

seashepherd.org

Worth Watching

Whale Wars (2008–2015)
Pirate for the Sea (2008)
At the Edge of the World (2008)
The Cove (2009)
Confessions of an Eco-Terrorist (2010)
Eco-Pirate: The Story of Paul Watson (2011)
Whale Wars: Operation Bluefin (2012)
Whale Wars: Viking Shores (2012)
Seal Wars (2012)
Defenders of the Wild – Ocean Raider (1993)
Black Harvest: The Fight for the Pilot Whale (1988)
Why Just One? (2016)
Ocean Warriors – Chasing the Thunder (2018)
Citizen Animal (2017)
Sea of Life (2017)
Sharkwater (2006)
www.sharktrust.org
Watson (2019)

---

[6] Quoted in Morris, D. B. (1995): *Earth Warrior: Overboard with Paul Watson and the Sea Shepherd Conservation Society*. Fulcrum Publishing, Golden, p. 115.

Part III
Heroes for the Environment

The Environmental Movement
Is Born – Rachel Carson and Silent Spring

After World War Two, the belief in technological progress was widespread. Many thought that it was possible for humanity to dominate, exploit and subdue nature without any negative consequence. This myth had a strong influence on how pesticides were used. Indeed, pesticides were expected to eradicate all insects, rodents, and microorganisms that were judged harmful or undesirable. Adverse side effects on the rest of nature (including humans) were not considered at all. Huge quantities of 'wonder chemicals' such as DDT were deployed in agricultural fields, settlements, and on human bodies, often with no safety precautions whatsoever. It was under these circumstances that Rachel Carson's story unfolded.

Rachel Carson, the American ecologist and writer, now a legendary character of the green movement, was born in Springdale, Pennsylvania in 1907. Influenced by her mother, she developed an interest in the natural world as a small child. She loved animals and enjoyed wandering in the woods. She asked her brother to give up hunting, because, as she put it, while hunting may be fun for the hunter, it can't be much fun for the hunted. Besides the living world, Rachel Carson's other passion was literature. She was only ten when her first short story was published, and she continued to publish stories and essays during the following years. Finally, she decided to study biology, and graduated as a zoologist from Johns Hopkins University.

From 1936 she worked for the US Bureau of Fisheries (the predecessor of the US Fish and Wildlife Service). She also wrote articles about nature for various newspapers and magazines. Her first book, *Under the Sea-Wind*, was published in 1941. Her next books, *The Sea Around Us* and *The Edge of the Sea*, were huge successes and real bestsellers. Released in 1953, the Oscar-winning documentary *The Sea Around Us* was based on Carson's book. All three parts of the trilogy are considered masterpieces on the natural history of the seas.

In 1956 Carson published a pioneering writing about environmental education. Entitled *Help Your Child to Wonder*, the article stresses the importance of discovering, experiencing, enjoying, and feeling nature – for children and adults alike. For nature has enormous power to improve, enrich, and heal the lives of those who are ready to take in the beauties of the living world. As she put it, 'Those who

© Springer Nature Switzerland AG 2019

L. Erdős, *Green Heroes*, https://doi.org/10.1007/978-3-030-31806-2_30

contemplate the beauty of the earth find reserves of strength that will endure as long as life lasts.'[1]

With *The Sea Around Us*, Rachel Carson became a well-known figure virtually overnight – at least in the United States. Worldwide fame came with *Silent Spring*.

The effects of pesticides already arouse Rachel Carson's attention in the 1940s. However, editors were uninterested, so she did not publish any writing about this topic until 1962. In that year, Carson's most influential book, *Silent Spring* was published. The title refers to the possible consequences of the indiscriminate use of chemicals. At that time, the phrase 'silent spring' seemed a real threat rather than a rhetorical exaggeration: in the 1950s and 1960s birds, together with other wildlife and companion animals died in masses all over the United States and other western countries. In *Silent Spring* Carson made use of her writing skills and scientific experience. The book documents how the misuse of pesticides harms humans and other organisms. Some pesticides persist in the soil, the water, and the bodies of living beings for a very long time. In addition, target pests are usually able to adapt to the chemicals, which means that ever larger quantities have to be used to maintain the desired effect. The natural enemies of the pests become rare or go extinct, so once the chemical treatment is over, pest populations are released from any control. Due to bio-accumulation and bio-magnification, even small amounts of chemicals can have fatal effects. Carson concluded her work with the following passage: 'The "control of nature" is a phrase conceived in arrogance, born of the Neanderthal age of biology and philosophy, when it was supposed that nature exists for the convenience of man. The concepts and practices of applied entomology for the most part date from that Stone Age of science. It is our alarming misfortune that so primitive a science has armed itself with the most modern and terrible weapons, and that in turning them against the insects it has also turned them against the earth.'[2] Carson did not deny that pesticides may be necessary in some cases, but she urged the care-

---

Bio-accumulation and Bio-magnification

Only if we have a correct understanding of ecological processes can we estimate the real effects of human interventions in nature. If we do not take into account bio-accumulation and bio-magnification, we may underestimate the risks associated with the overuse of synthetic pesticides. Bio-accumulation means that pesticides, after entering our bodies, tend to accumulate in lipid-rich tissues such as body fat, breast, and bone marrow. Pesticides may remain in these tissues for a very long time. The longer the lifespan of a living being is, the more pesticides may accumulate. Bio-magnification means that the concentration of pesticides increases as these travel up the food chain from plants to herbivores to predators. Even if pesticide concentration is relatively low in plants, concentration will be higher in herbivores and still higher in predators. Thus, long-lived predatory animals are particularly seriously affected by the indiscriminate use of chemical pesticides.

---

[1] Carson, R. (1956): Help Your Child to Wonder. Woman's Home Companion, July 1956, pp. 24–48.
[2] Carson, R. (2002): Silent Spring. Houghton Mifflin Company, Boston, p. 297.

ful use of chemicals, and offered solutions that are cheaper, easier, more effective, and less harmful than the irresponsible use of dangerous chemicals.

Although her views were clearly supported by scientific facts, Carson received heavy criticism from the chemical industry. They published parodies and tried to discredit Carson, depicting her as an unscientific and hysterical woman and a communist sympathiser. Despite the unfair attacks, however, *Silent Spring* was an immediate success. Environmental awareness increased rapidly, and citizens started to demonstrate against the destruction of the living world. Major environmental laws were passed, and the USA quickly became a leader in international environmental issues – this momentum lasted till the 1980s. The modern environmental movement was born, and there can be no doubt that Rachel Carson's *Silent Spring* played a major role in this process.

In 1962 Carson received the Albert Schweitzer Medal of the Animal Welfare Institute for contributing to the protection of animals from deadly pesticides. In 1963 she was awarded the Audubon Medal of the National Audubon Society. Rachel Carson died of cancer in 1964. She was awarded the Presidential Medal of Freedom, the highest civilian award of the United States, posthumously in 1980.

Silent Spring has been translated into about 30 languages, and is considered one of the most influential books of the twentieth century. It inspired a large number of environmentalists, including Al Gore, Denis Hayes, Arne Naess, and David Suzuki. Rachel Carson changed the way we think about chemical pesticides, and, more generally, about human interventions into nature. Synthetic pesticides are poisons: sprayed to the field, they destroy the natural world and then come back to us. Their use has to be minimised, and, if possible, avoided completely. Messing with nature may well be more dangerous than most pests we would like to eradicate. The reception of *Silent Spring* taught us that industrial forces are ready to use the most nefarious methods to keep profits high. However, standing up against them is possible, as has been shown so many times since the 1960s. After *Silent Spring*, environmental organisations became popular, and environmental problems could not be ignored any more in politics. It is clear, therefore, that *Silent Spring* represents a turning point. Unfortunately, however, environmentalism continues to be a marginal topic. Although there are some successes, global politics and economy are mostly shortsighted. Thus a new turning point is urgently needed to complete what Rachel Carson started.

'Have we fallen into a mesmerized state that makes us accept as inevitable that which is inferior or detrimental, as though having lost the will or the vision to demand that which is good?'[3] Carson contemplated in *Silent Spring*. The situation may not be entirely hopeless. By growing our own vegetables or buying ecologically grown products we can prove that we do have a vision of a future that is not poisoned.

---

[3] Carson, R. (2002): Silent Spring. Houghton Mifflin Company, Boston, p. 12.

Worth Reading

Carson, R. (1991). *The sea around us*. Oxford: Oxford University Press.
Carson, R. (1998). *The edge of the sea*. Boston: Houghton Mifflin Company.
Carson, R. (2002). *Silent spring*. Boston: Houghton Mifflin Company.
Carson, R. (2007). *Under the sea-wind*. New York: Penguin Books.

Worth Browsing

www.rachelcarson.org

Worth Watching

The Sea Around Us (1953)
A Sense of Wonder (2008)

Environmentalism Gaining Momentum – Denis Hayes and Earth Day

During the 1960s, environmental awareness grew considerably, not least because of Rachel Carson's *Silent Spring*. This process culminated in the first large-scale green event of the world: on April the 22nd, 1970, an estimated 20 million people participated in Earth Day. It was by far the largest public demonstration in the history of the United States. Besides historical antecedents and circumstances, the talents of Denis Hayes, the principal organiser of Earth Day undoubtedly played an important role in this enormous success.

Denis Hayes was born in 1944. He spent much of his childhood in the town of Camas, where he had the opportunity to experience the effects of pollution first-hand. There was a papermill near their home, and, due to sulphur dioxide and hydrogen sulphide flowing from the chimneys, the small boy had a sore throat each morning. As a student he spent some time backpacking. He was 21 when, on a hitchhiking tour in Africa, he understood that human society should co-operate with nature rather than fight against the principles of ecology. It was this moment when he decided to do his best to support this transition. After returning home and becoming a university student, he was ready to change the world. He only needed a great idea.

The idea of Earth Day was first proposed by peace activist John McConnell in 1969. He suggested that there should be a holiday celebrating the living systems of planet Earth. The project was embraced by politicians, scientists, and the city of San Francisco, where the first Earth Day was organised on March the 21st, 1970. But the history of Earth Day is not that simple: it also has a second origin, which has its roots in the mind of Gaylord Nelson.

Democrat Senator Gaylord Nelson had always had an interest in environmental issues. Upset by the terrible consequences of the 1969 Santa Barbara oil spill, he decided to organise environmental teach-ins across the US. He teamed up with Republican Congressman Paul McCloskey, and they started to seek organisers for the event. Denis Hayes planned to volunteer as a local coordinator, but soon found himself in charge of the whole campaign. He had two basic ideas on how to organise the event, and both turned out to be decisive in the final success. First, he suggested

© Springer Nature Switzerland AG 2019
L. Erdős, *Green Heroes*, https://doi.org/10.1007/978-3-030-31806-2_31

that not only universities and colleges, but also primary and secondary schools and local communities should be involved. Second, he had a vision of uniting various environmental branches and organisations under a common flag. Wilderness preservation, human health, endangered species, water pollution, and nuclear tests were largely viewed as unrelated issues, but were brought together on Earth Day.

After the first Earth Day in San Francisco in March 1970, the second, US-wide Earth Day took place on April the 22nd, 1970. Denis Hayes and his team did a great job: about 10,000 public schools, more than 2000 universities and colleges, and thousands of different groups and communities participated. People planted trees, collected rubbish, attended lectures, demonstrations, and concerts. The event attracted hundreds of thousands of citizens in New York City alone, where even Fifth Avenue was closed for the rally. A lot of famous musicians, actors, and politicians joined the cause. Earth Day received considerable media attention: environmental concerns became a front-page issue.

Earth Day 1970 had a massive effect on both the public and politics. Major environmental laws were passed that would have been unthinkable a few years earlier: the Clean Air Act, the Clean Water Act, and the Endangered Species Act, to name just a few. In 1990 Hayes returned to direct Earth Day, which was developed into a global event. In that year there were some 200 million participants in 141 countries. Today Earth Day is the largest secular holiday of the world.

Green Holidays

Although Earth Day is the best-known green holiday, by no means is it the only one. Similar holidays include World Water Day (March the 22nd), World Day for Laboratory Animals (April the 24th), World No Tobacco Day (May the 31st), World Environment Day (June the 5th), Clean Up the World Weekend (on one weekend in September), Car Free Day (September the 22nd), World Animal Day (October the 4th), and Buy Nothing Day (the last Friday or Saturday in November).

Although Denis Hayes jokingly remarks that he peaked at the age of 25 as the coordinator of Earth Day, it is clear that he has had many other significant achievements. Hayes was only 35 when he was appointed as director of the Solar Energy Research Institute but resigned when the funding was cut during the Reagan administration. He also served as senior fellow at the Worldwatch Institute, adjunct professor of engineering and human biology at Stanford University, Regents' Professor at the University of California, and board chair of Earth Day Network. From 1992 he has been the president of the Bullitt Foundation, an organisation promoting sustainability in the Pacific Northwest of the US. With the Bullitt Center, the greenest commercial building of the world, the Foundation has demonstrated that it is possible to construct and operate buildings in an environmentally friendly way. The Bullitt Center has geothermal heating and cooling, by means of solar panels it produces more electricity than it needs, runs its own composting toilet system, collects rainwater, cleans greywater in a roof garden, has a bicycle garage, and applies a number of energy saving solutions. During construction, every effort was made to reduce carbon footprint and to avoid unhealthy materials. The whole design mimics the functioning of a natural ecosystem.

Denis Hayes' main area of interest is renewable energy. This was the focus of his 1977 book *Rays of Hope*, which provides readers with a comprehensive overview of energy issues. Although some facts and ideas of the book are clearly outdated, the main message is as timely now as it was 40 years ago. Hayes argues that sooner or later we have to switch from fossil fuels to more sustainable sources, simply because we will run out of coal, oil, and gas. Of course, due to the terrible environmental and health effects of fossil fuels, the switch should occur sooner rather than later. Hayes is convinced that the transition is possible with a combination of using less energy and relying on renewable sources. Using less energy requires some technological improvements (for example, insulating our homes) and the rationalisation of our lifestyles (much of our consumption is only some kind of status symbol, but not really necessary for an acceptable quality of life). Among renewable sources, Hayes considers solar energy the most promising, although other renewables may also play important roles regionally. Hayes also discusses the possibilities and limitations of nuclear energy, noting that it is risky, expensive, and the waste disposal is unsolved – characteristics that have not changed much during the 40 years since Hayes' book was published. In contrast, solar energy is safe and getting cheaper each year. Why, then, has the transition not happened yet? According to Hayes, since we do have the technology, any delay has political rather than technical reasons. As he put it, 'The jingle of the cash register can drown out the voices of the unborn.'[1]

Denis Hayes' second book, *The Official Earth Day Guide to Planet Repair*, was published in 2000. It is a concise and enjoyable work on global warming, and it informs readers about how they can take action to make the world a better place. Hayes points to the fact that energy issues lie at the heart of many environmental and social problems. Using fossil sources not 'only' causes global warming, but is also responsible for oil spills and smog, and contributes to acid rain, respiratory diseases, and foreign economic dependency. This, however, also means that moving toward renewable energy can solve multiple problems simultaneously. This would be a win-win situation, except for a small elite that makes profit by destroying the planet. Hayes suggests that solving global problems should start at many levels, from our households up to international agreements. And the reader is encouraged to reduce their ecological impact and to urge politicians to protect the environment.

With his wife Gail, Denis Hayes published his most recent book in 2016. *Cowed* describes the negative ecological impacts meat and dairy products have in the form of immense soil, water and air pollution. Factory farms are terrible for the animals and human health alike. The book is a convincing argument for reducing meat consumption and buying organic products – for the sake of humans, animals, and the environment.

---

[1] Hayes, D. (1977): Rays of Hope: The Transition to a Post-petroleum World. W. W. Norton & Company, New York.

Denis Hayes is one of the most respected environmentalists. He received the Jefferson Medal of the American Institute for Public Service for the greatest public service by an American under 35. He was selected as one of the 100 most influential conservation figures of the twentieth century by the National Audubon Society, and as Hero of the Planet by *Time* Magazine. Hayes Freedom High School and Denis Hayes Street in his native Camas are named after him.

For the last 50 years, Denis Hayes has been a leading character of environmentalism. But if he is asked whether he is satisfied with what has been achieved since the first Earth Day in 1970, the response is only partly positive. It is undeniable that humans' attitude toward the environment has changed profoundly. People are ready to stand up for a clean environment. Everyone knows that human health depends on the health of the environment. However, little progress has been made at the global scale. Huge areas of forest are cleared every year, greenhouse gases are building up in the atmosphere at an unprecedented rate, plastic garbage is accumulating in the oceans, the diversity of life is decreasing in what is considered a mass extinction event, and the list goes on and on. Neither Earth Day nor any other event has been able to reverse these trends.

Our species continues to rape, ruin, degrade and devastate nature all year round. It is nice that we have one day to celebrate our planet. Of course, real change would need that we become more environmentally conscious each day, constantly striving to minimise our environmental impacts. Let's make every day Earth Day! This is the real message of Earth Day.

Worth Reading

Hayes, D. (1977). *Rays of hope: The transition to a post-petroleum world.* New York: W. W. Norton & Company.
Hayes, D. (2000). *The official earth day guide to planet repair.* Washington, DC: Island Press.
Hayes, D., & Hayes, G. B. (2015). *Cowed: The hidden impact of 93 million cows on America's health, economy, politics, culture, and environment.* New York: W. W. Norton & Company.

Worth Browsing

www.bullitt.org
www.bullittcenter.org
www.earthday.org

Worth Watching

Earth Days (2009)
Carbon Nation (2010)

The Greenpeace-Story

Greenpeace is the largest and one of the most respected environmental organisations in the world. But its story started with a handful of peace activists who gathered around the kitchen table and made plans to save the world.

During the 1960s, as a response to the Vietnam War, the Cold War, and the arms race, the peace movement gained strength. Anti-war protests were linked to the demonstrations against nuclear weapons tests. The peace movement was especially powerful in Vancouver, a city in Southwest Canada. Some of the protesters became friends and met regularly, creating an informal group which would later become Greenpeace. The most dedicated members were, among others, Robert Hunter, Jim and Marie Bohlen, Ben Metcalfe, Bill Darnell, Paul Cote, Terry Simmons, Bob Cummings, and Irwing and Dorothy Stowe. Irwing Stowe would always greet the other members by flashing the 'V' sign and saying 'peace.' On one occasion, Bill Darnell added, 'Make it a green peace.'[1] This is how the name 'Greenpeace' was born.

The first mission of the group was to prevent the US from carrying out a nuclear bomb test at Amchitka Island off the Alaskan coast. Everyone agreed on the goal of the project, but nobody quite knew how to achieve it. Finally they accepted Marie Bohlen's proposal to sail a boat to Amchitka, as the military would not explode the bomb if there was a boat full of civilians.

However, they had no boat, neither had they money to buy or charter one. So they organised a benefit concert. Although no one really expected, the concert turned out to be very successful, so the financial background was secured. The next task was to find someone willing to charter a boat that was to sail to a nuclear test zone. Captain John Cormack agreed to steer his fishing boat to Amchitka. The team started their voyage on September the 15th, 1971. The unusual enterprise quickly got in the focus of media interest, and many followed the news about the courageous activists. The team was ordered away from Amchitka by the US Coast Guard. Also, as the test

---

[1] Quoted in Weyler, R. (2004): Greenpeace: How a Group of Ecologists, Journalists, and Visionaries Changed the World. Rodale, Emmaus, p. 67.

© Springer Nature Switzerland AG 2019

L. Erdős, *Green Heroes*, https://doi.org/10.1007/978-3-030-31806-2_32

was postponed and winter was approaching, the weather became too harsh for the small fishing boat, which returned to Canada. The hydrogen bomb was detonated on November the 6th, causing a massive natural destruction and radioactive pollution. Greenpeace seemingly lost the battle, but it was not a total failure. The organisation became well-known, and their cause was supported by millions.

Over the next period, the activity against nuclear tests remained in the focus of Greenpeace. In 1973 Greenpeace activists approached the Moruroa Atoll, where France was conducting nuclear tests. The French navy raided the boat of Greenpeace and brutally attacked the unarmed environmentalists. One of the activists, David McTaggart suffered serious injuries and narrowly escaped being blinded. The French navy came up with a story claiming that McTaggart had slipped on the deck, but they were not aware of the fact that photographs documented the brutal attack. The photos were published and provoked a global outrage, which was extremely embarrassing for the French navy. In 1974 France announced that it would end its atmospheric nuclear testing programme.

The scope of Greenpeace widened when young members suggested taking action for nature and animals. Paul Watson, Robert Hunter, and Paul and Linda Spong were instrumental in launching Greenpeace's anti-whaling campaign. The activists steered small inflatable boats between the whaling vessels and the whales, trying to protect the animals. They managed to save hundreds of whales each year and won the sympathy of the public. Greenpeace played an important role in changing the whaling policy of Australia, which became a fierce opponent of whaling. Moreover, partly as a result of the efforts of Greenpeace and other organisations, the International Whaling Commission banned commercial whaling completely in 1986.

The anti-sealing project of Greenpeace started in 1976. Paul Watson and Robert Hunter did not hesitate to risk their own lives for baby seals. In 1977, movie star and animal advocate Brigitte Bardot, accompanied by dozens of journalists, joined Greenpeace activists on the ice fields and hugged a seal pup. A few years later the European Community banned the import of certain seal products. In 1978 Patrick Moore, an ecologist of Greenpeace spent his third sealing season on the ice sheets. Upset by the blood, the dead seal cubs, and the mother seals desperately looking for their babies, he embraced a baby seal and asked the hunters no to kill that animal. 'I want this one pup to live. For 3 years we've watched the sealers come through here. They leave nothing but carcasses behind. This is a nursery, for God's sake. Just this one. For all the people of the world who have demonstrated their opposition to this hunt,'[2] Moore argued. But there was no mercy. Canadian policemen arrested Moore, and the helpless animal was slaughtered. In a few seconds the baby seal was nothing more than a bleeding, skinned carcass. Moore was brought to jail for violating the Seal Protection Act, according to which nobody is allowed to approach seal pups. Nobody, that is, except seal hunters...

---

[2] Quoted in Weyler, R. (2004): Greenpeace: How a Group of Ecologists, Journalists, and Visionaries Changed the World. Rodale, Emmaus, p. 499.

There was a time when some countries obviously thought that the best way of radioactive waste management was to throw the nuclear material into the ocean. Greenpeace could not remain silent: they informed the public about the scandalous practice. Also, Greenpeace rubber boats approached the ships that carried nuclear waste and tried to prevent waste being dumped in the water. On one occasion a Dutch ship dropped the barrels full of nuclear waste onto a rubber boat, thereby directly threatening the lives of the protesters; it was just good luck that nobody died. Thanks to the environmentalists' efforts, this type of radioactive waste disposal was banned.

Some of the campaigns may have been less spectacular, but they nevertheless played an important role in improving the quality of our environment. Greenpeace demanded that sewage should not be discharged into rivers without purification. The environmental organisation also worked tirelessly to persuade governments and corporations to reduce the air pollution of power plants. As a result, the air and the rivers of many European countries are much cleaner now than they were a few decades ago.

Nuclear weapons tests destroyed heavenly places, turning them into uninhabitable areas. Indigenous people were treated as experimental subjects in a huge laboratory. Inhabitants of the Rongelap Atoll (Marshall Islands, Pacific Ocean) were suffering from tumours and birth defects caused by nuclear tests. In 1985 they asked Greenpeace to evacuate them. The relocation was carried out by the Rainbow Warrior, the flagship of Greenpeace. The next mission of the crew would have been a demonstration against French nuclear tests. However, events took an unexpected turn. On July the 10th, 1985, at 11:50 p.m., a bomb detonated on the Rainbow Warrior. The ship was sunk by French secret agents in an attack called Operation Satanic. Fernando Pereira, a 35-year-old member of the Greenpeace crew, drowned during the incident. Initially France denied responsibility, but later, as evidence accumulated, they officially admitted that it had been a French secret operation.

In the 1980s some countries and corporations started to develop plans to exploit the resources of Antarctica. At the same time, numerous environmental organisations, including Greenpeace, demanded that Antarctica, the last continent that is relatively free from major human destructions, should be protected. France turned out to be an unlikely ally of Greenpeace, because the new prime minister wanted to repair the damage caused by the public relations disaster of the Rainbow Warrior affair. The efforts of the environmentalists succeeded in 1991, when the Madrid Protocol was signed, prohibiting mining on Antarctica.

In the 1990s, Greenpeace's campaign against Shell gained international media attention. The oil company planned to sink one of its old oil platforms into the Atlantic Ocean. Quite naturally, Greenpeace started a protest, demanding that the platform be dismantled on land in an environmentally responsible way. However, Shell preferred simply dumping the structure into the ocean, as this would have cost the company less. Greenpeace activists occupied the abandoned building and barricaded themselves. After the aggression of Shell security guards had got news coverage throughout the world, a boycott started to unfold. Eventually Shell decided to decommission the platform on land. Today this success is seen by many as a symbol of what environmental organisations can achieve if they are supported by a large number of conscientious consumers.

Around the millennium Greenpeace refined its strategy to meet the demands of the twenty-first century, without compromising its original vision and principles. The creative use of social media in delivering short but efficient messages to a wide audience is of particular importance. Also, increasing emphasis is being placed on the connections between environmental issues and social justice.

Though Greenpeace is best known for the organisation's spectacular actions, their activity is actually much wider, and includes environmental research, education, and lobbying. Greenpeace has been involved in countless environmental issues such as protecting the ozone layer, designating nature reserves, fighting deforestation, protecting human health, and mitigating the effects of oil spills, to name just a few. The efforts of Greenpeace are recognised throughout the world and are supported by millions of people. In 2002, Nobel peace prize laureate Desmond Tutu visited Esperanza, one of Greenpeace's ships, he blessed the crew, and assured them of his full support in their fight for a clean, peaceful and nuclear-free future.

Greenpeace is usually criticised for being sensationalist and placing too much emphasis on radical demonstrations. However, it should be kept in mind that protests are always preceded by detailed studies and negotiations. If corporations or authorities cannot be convinced that they should act responsibly, Greenpeace activists have no choice but to attract media attention with impressive, sometimes even risky protests. There is no doubt that the Earth would be in a much worse condition without the dedicated work of Greenpeace activists. Some critics claim that Greenpeace simply keeps hindering development efforts. This is not true, as Greenpeace always offers alternative routes that enable development in a clean and safe way.

Greenpeace does not accept corporate or government donations thereby guaranteeing that it cannot be influenced by corporate or political interest. Instead, the organisation relies on individual supporters.

Every single achievement of Greenpeace is a success for the environment, the Earth, human health, and our common future. Greenpeace is working for all of us, as do other green organisations. But when will this struggle end? According to Kumi Naidoo, who headed Greenpeace from 2009 to 2015, 'ultimate success will be achieved when we are no longer necessary.'[3]

Worth Reading

Erwood, S. (Ed.). (2011). *The Greenpeace Chronicles: 40 years of protecting the planet.* Amsterdam: Greenpeace International.
Brown, M., & May, J. (1989). *The Greenpeace story.* London: Dorling Kindersley.
Weyler, R. (2004). *Greenpeace: How a group of ecologists, journalists, and visionaries changed the world.* Emmaus: Rodale.

---

[3] Naidoo, K. (2011): Greenpeace's 40 years of activism prepare us for our greatest threat. The Guardian, September the 15th, 2011. www.theguardian.com/environment/2011/sep/15/greenpeace-40-years-activism

Worth Browsing

www.greenpeace.org

Worth Watching

The Rainbow Warrior (1993)
Greenpeace: Making a Stand (2006)
Greenpeace: The Story (2011)
Greenpeace: From Hippies to Lobbyists (2011)

Chico Mendes – A Martyr for the Rainforest

Tropical rainforests contain a baffling biodiversity and are essential for the health of the biosphere, yet these highly complex ecosystems are being devastated at an alarming rate. Deforestation has serious global consequences and it also presents an immediate threat to human populations whose livelihoods depend on the forest. In Amazonia, both Native Americans and rubber tappers have been exploiting the goods and services of the jungle in a sustainable manner. Their desperate fight to protect their lifestyles is perfectly demonstrated by the story of a tragic hero, Chico Mendes.

Why Are Tropical Rainforests Important?
Rainforests have an outstanding ecological importance. They cover some 6% of the Earth's terrestrial surface yet provide habitats for an estimated 50% of all species. One hectare of rainforest is home to more tree species than the whole of Europe. Of the species living in the world's rainforests, many could be used as medicines or for nutrition, provided that we do not eradicate them. Rainforests sequestrate immense amounts of carbon, thereby mitigating global warming. Even today, many people live in and from the rainforests; their survival and culture depend on the existence of these ecosystems. Rainforests protect the soil from erosion and prevent landslides and floods. Also, rainforests play a major role in climate regulation, both regionally and globally.

Tropical deforestation has an estimated rate of 100,000 km^2 per year, which translates to a rainforest loss of ca. 190,000 m^2 each minute. Day and night, all year round. In the not-too-distant future, if the current rate continues, no rainforests will remain outside nature reserves. In addition, a considerable proportion of rainforests within reserves will also be gone, as reserves usually exist only on paper.

© Springer Nature Switzerland AG 2019
L. Erdős, *Green Heroes*, https://doi.org/10.1007/978-3-030-31806-2_33

Chico Mendes was born in 1944 in the jungle of the northwest Brazilian state of Acre. He was a son of a rubber tapper and had a very hard childhood indeed: he received no formal education but had to help his parents instead. As a small kid, Chico Mendes collected firewood and carried water from the river. Later he accompanied his father on his latex gathering routes and helped him with smoking latex; they had to work up to 15 h a day. The family grew vegetables in their garden, and they hunted wild animals and collected fruits in the forest. Unlike most rubber tappers at that time, Chico's father was literate and taught his son to read and write.

Chico Mendes was almost completely isolated from the outside world until the age of 18, when his family was visited by a mysterious man named Euclides Fernandes Távora. A leftist thinker and former military officer, Távora was on the run because of his former political and guerrilla activity. The two had lengthy discussions about the political situation and the oppression of rubber tappers. It was Távora who instilled in Chico the idea of organising workers' unions among the rubber tappers.

Rubber tappers were treated almost like slaves. They were denied education and health care, and were ruthlessly exploited by rubber merchants, who found it easy to cheat the illiterate rubber tappers. And then, as if the oppression and enslavement were not bad enough, another major problem emerged, which threatened the very existence of rubber tappers. From the 1960s the large-scale deforestation of the Amazon Basin became a priority for the government of Brazil. The rainforests were intended to absorb inhabitants from the overpopulated southern and eastern regions of the country. Anyone who could prove that they used the land productively took possession of that land. Wealthy ranchers wanted to take advantage of this opportunity: they moved to the sparsely populated rainforest areas, cleared the forests, and converted them to pastures. There was only one restriction: areas inhabited by Native Americans or rubber tappers could not be occupied. The solution to this restriction, however, was simple: ranchers usually hired gunmen to chase the inhabitants away or kill them. Hundreds were murdered, but in only a handful of cases were the perpetrators convicted.

Some of the ranchers were criminals with extremely long rap sheets. They came to west Amazonia, a land beyond the reach of the law, to avoid being arrested. One of the most dangerous of these outlaws was Darly Alves da Silva. When he settled in western Amazonia in 1974, he had already been responsible for the deaths of several people. He lived on his huge ranch with his wife, some lovers, about 30 children, and a small private army. The family made their living by smuggling drug and gun, but they probably also did contract killing. To make things worse, the clan had excellent relations with the local police chief.

Mendes started to fight for better working conditions when he was in his twenties. He wrote letters to different authorities describing the rubber tappers' terrible situation. Also, he taught rubber tappers to read and write. He took part in the organisation of a workers' union, recruiting a considerable number of new members. Later, he was elected council member of the local municipality. But his most enduring achievements are connected to rainforest protection.

Rubber tappers' resistance against deforestation started to unfold in the 1970s, when groups of rubber tappers went to the camps of the loggers, blocked logging operations with their bodies, and ordered the loggers to leave. The reactions were extreme. Police usually intervened violently, and ranchers' responses to the blockades were deadly. Anyone who dared to confront ranchers took serious risks: ordinary rubber tappers and union leaders, but even lawyers, priests, and bishops who supported the movement received death threats, were beaten or killed. Chico Mendes was no exception. As he was particularly successful in gathering participants for the blockades, he quickly became one of the ranchers' targets.

In 1985 Mendes organised the first national conference of rubber tappers, an event that attracted many politicians and environmentalists, and received considerable media coverage. When rubber tappers joined forces with environmental groups and Native Americans, Mendes quickly became an internationally recognised figure. Supported by environmentalist groups, he travelled to the US, where he negotiated with representatives of the World Bank and the Inter-American Development Bank. Mendes explained them that development projects supported by international banks typically result in rainforest destruction and human rights abuse. He suggested that projects should meet certain environmental and social requirements to prevent or minimise further harm. In 1987 Chico Mendes received the UN Global 500 Award and the Environment Medal of the Better World Society.

Despite his growing fame, Chico Mendes remained a modest man. He devoted most of his time to organisational work, which reduced his income. His financial situation was miserable: he often had to hitchhike because he couldn't even afford to buy a bus ticket. And he could wear a suit on meetings only because he received one from the Catholic Church.

Meanwhile he continued with the blockades, which achieved some remarkable successes: they managed to save large rainforest areas from being cleared. Mendes also elaborated the concept of extractive reserves, an economically viable alternative to rainforest clearing. Extractive reserves are protected areas where only sustainable activities are allowed, such as the extraction of latex, and the collection of fruits and medicinal plants. The model was able to integrate ecosystem preservation, social justice, and economic development. In 1988, based on the idea of Mendes, the first extractive reserve was established.

As Mendes gained international recognition and became more and more effective in protecting the forest, the ranchers' hatred towards him started to increase drastically. They tried to bribe Mendes, but he was absolutely incorruptible. So he received death threats. From that point on, Chico Mendes was not just one of the ranchers' enemies, but he was the primary target. He was now at the top of the death list. The situation, however, deteriorated even further.

A blockade organised by Mendes prevented the Alves family from clearing a forest area. Though Alves received compensation from the state, he was furious. Darly Alves da Silva was not the kind of man who tolerated being defeated.

The final straw was when Mendes, with the help of a lawyer whom he had met at a conference, exposed the criminal record of the Alves clan. An arrest warrant was issued for Darly Alves. However, the police did nothing for a surprisingly long period, partly because they did not want to ruin the good connections with Alves, and partly because they were afraid of him. By the time the military police set off a commando to arrest him, Alves had been informed and escaped. Forced to hide, Alves vowed to take revenge. He declared that Chico Mendes would die before New Year's Eve.

Chico Mendes was marked for death. He and his family were living in constant fear. Mendes wrote to several authorities about the death threats, and he received two guards from the military police. On December the 22nd, 1988, Mendes was at home with his wife, their two children, and two guards. In the evening Mendes stepped out to the backyard and was shot by an assassin who had been waiting there.

The bandits had no doubt they could murder Chico Mendes with impunity. However, this time they were wrong. The act provoked a public outcry and authorities came under heavy international pressure to punish the killers. Darly Alves da Silva, the mastermind behind the murder was arrested, together with his son Darci, who pulled the trigger. They were sentenced to 19 years in prison. They were released after serving the third of it. There was strong evidence indicating that other ranchers and even some politicians and police officers had been involved in the crime, but they have not been called to account.

The rubber tappers lost their charismatic leader, but the spirit of Chico Mendes lives on. Many have followed in his footsteps. Several of his colleagues and friends became environmentalists and politicians and were instrumental in changing Brazil's forest policy. Today there is a network of extractive reserves in Brazil. The courage of Chico Mendes earned him a folk hero status, and his figure appeared in pop culture. Paul McCartney dedicated his song *How Many People* to the memory of Chico Mendes. *The Burning Season*, a 1994 film is based on the life of Mendes.

'My dream is to see this entire forest conserved because we know that it can guarantee the future of all people who live in it,'[1] Mendes said. He showed that the rainforest is able to provide livelihoods for millions of people, provided that it is not cleared for a few thousand ranchers and speculators. Unfortunately, the clearcutting of the jungle continues in Amazonia and elsewhere. And the fight for the forest is as dangerous as it was in Mendes' time. Among those who want to protect the forest, thousands have received death threats and dozens were killed during the last couple of years alone. Ecology has taught us that tropical rainforests have a global importance in climate regulation and other ecosystem services. Therefore, we would do well to take immediate action for rainforests, if we want to preserve a liveable planet.

---

[1] Mendes, C.; Gross, T. (1989): Fight for the Forest: Chico Mendes in His Own Words. Latin American Bureau, London, p. 6.

Worth Reading

Mendes, C., & Gross, T. (1989). *Fight for the forest: Chico Mendes in his own words*. London: Latin American Bureau.

Revkin, A. (2004). *The burning season: The murder of Chico Mendes and the fight for the Amazon rain forest*. Washington, DC: Island Press.

Rodrigues, G. (2007). *Walking the forest with Chico Mendes: Struggle for justice in the Amazon*. Austin: University of Texas Press.

Shoumatoff, A. (1990). *The world is burning*. Boston: Little, Brown and Company.

Worth Watching

Voice of the Amazon (1989)
The Burning Season (1994)

Missing in Action – The Story of Bruno Manser

Similar problems often induce similar reactions irrespective of the geographic location. If greed and corruption result in rainforest destruction and human rights violation, heroic figures may naturally emerge and tragic stories may unfold, in Borneo just like in Amazonia.

Bruno Manser was born in 1954 in Basel, Switzerland. Like many children, he was interested in the natural world: he walked in the forests, observed the behaviour of animals, read books about biology, and particularly enjoyed climbing trees. But unlike most usual kids, he also seemed to be interested in the tougher side of life. Using twigs and leaves, he would prepare a 'bed' on the balcony, where he would spend the night even in wintertime. He learned how to sew and knit, he carved buttons and made shoes, and could prepare meals. His dream was to become a scientist who lives in the jungle and studies the natural world. In one of his school essays, 15-year-old Bruno wrote 'Freedom can only be enjoyed in nature – not in the world of technology.'[1] After finishing secondary school, Manser studied medicine, but he soon lost interest. In 1973 he refused compulsory military service, for which he was sentenced to 4 months' imprisonment.

After his release from jail, Manser started his journey to become the person he wanted to be. He enrolled in an agricultural school to learn how to herd animals. Accompanied by his brother Erich and his dog Prinz, Bruno Manser spent ten summers in alpine pastures. It was a life as close to nature as was possible in a western European country. They got up well before sunrise and spent the day looking after the animals, milking, and making butter and cheese. When the daily chores were done, they usually went on excursions, visited locals, or met other herdsmen.

In the early 1980s, Manser started to nurse new plans: he wanted to visit a nomadic people that did not use money and lived self-sufficiently in nature. The Penan tribe seemed to be a perfect choice.

---

[1] Quoted in Suter, R. (2015): Rainforest Hero: The Life and Death of Bruno Manser. Bergli Books, Basel, p. 38.

© Springer Nature Switzerland AG 2019
L. Erdős, *Green Heroes*, https://doi.org/10.1007/978-3-030-31806-2_34

The rainforests of Borneo were inhabited by several indigenous peoples, including Penan. Although, as a result of mistaken projects, virtually all Penan have settled by now, they had a nomadic hunter-gatherer lifestyle originally. Penan lived in small groups in the forests of the Malaysian federal state Sarawak. They hunted with blowpipes, fished, and made porridge from sago palm. They used the plants as medicines, clothing, and building material for their temporary settlements. When the sago palms of a given site were harvested, the group moved to a new locality, allowing the jungle to regenerate. The Penan people possessed a rich culture, their own language, religion, and traditions. This culture became threatened by intensive deforestation in the second half of the twentieth century.

Logging destroys the forest, the very home of the Penan tribe. Animals disappear, rivers and brooks get muddy because of erosion, and fish become rare. To put it simply, Penan become homeless overnight. Of course, logging brings a lot of money, especially if it is followed by planting oil palms. However, the profit goes to a handful of timber barons and corrupt politicians, who are able to accumulate fabulous wealth. Despite some promises, and despite the incredible amount of foreign currency flowing in from unsustainable timber export, ordinary citizens' quality of life did not improve, while people living in the forest were denied their most basic rights.

Manser arrived in Borneo in 1984. First he joined a cave exploring expedition, then he walked into the jungle to find the Penan people. He was alone, carrying a heavy backpack, equipped with a bush knife, a compass, and a map. He got lost, ran out of food and water, but eventually met two Penan and followed them to their village. A peaceful people, Penan accepted the stranger. Manser immediately started to learn their language, and got to know how to prepare food, make tools, use the blowpipe, and identify medicinal plants. Soon Manser adopted the appearance and behaviour of his new friends. The Penan called Manser 'Laki Penan', 'the Penan Man', indicating that he really became one of them. He was almost indistinguishable from the Penan, except for his glasses and notebooks. Manser observed the plants, the animals, the people and their objects, and recorded everything that seemed important or interesting. He also made excellent drawings and paintings.

By the time Manser arrived in Borneo, opposition to deforestation was already happening. But the protests of the indigenous people suffered from several weaknesses. They faced powerful corporations backed by the Malay authorities. They lacked any land rights. Moreover, the natives were offered tiny compensations, and new schools and hospitals were promised if they accepted deforestation. While some tribes gave in, Penan tended to refuse compensations or developments. It quickly became clear to Manser that his new life in the rainforest paradise was under siege. He had to realise that it was impossible to escape industrial civilisation. In his diary he asked himself, 'Does this paradise really have to die and make way for chainsaws and bulldozers?'[2] At first Manser was unsure whether he should be involved in the protests, but he decided to do his best to help his Penan friends.

---

[2] Quoted in Straumann, L. (2014): Money Logging: On the Trail of the Asian Timber Mafia. Bergli Books, Basel, p. 118.

Manser tried to draw international attention to rainforest destruction, so he invited journalists to Borneo. In 1986, *Geo*, a respected German popular science magazine published an article about the situation of the Penan tribe. After a while, the Swiss Tarzan and his crusade made news in countless other newspapers and magazines. On behalf of tribal chiefs, Manser officially requested that the government should designate a protected area. He also helped Penan to organise demonstrations and set up blockades on logging roads. Though protests remained peaceful, they were not tolerated by the authorities: blockades were regularly followed by the imprisonment, torture, and humiliation of participants and other environmental activists who opposed logging. The fight caught the attention of renowned environmental and human rights organisations, which started to urge for a boycott on Malaysian timber.

Manser was declared enemy of the state. Police arrested him, but Manser managed to run away. The escape resembled an action scene from a Hollywood movie: when the police car stopped to refuel, Manser jumped into the jungle. Policemen shot after him but were unable to follow Manser in the dense undergrowth. In the following years police made serious efforts to capture the troublemaker. They forced Mutang Urud, one of Manser's closest allies to inform them about the whereabouts of Manser. Mutang Urud grew up in a Kelabit village in the rainforest. He earned a university degree and worked as a manager, but then returned to his tribe to support them in resisting deforestation. When police approached him, he pretended to co-operate but he did not betray his friend. A few years later Mutang Urud fled to Canada, where he was granted political asylum. Simultaneously with the efforts to arrest Manser, Malay authorities also tried to discredit him, depicting Manser as a person who wants to oppress indigenous people and keep them in poverty.

After 6 years in the jungle, Bruno Manser was informed that both of his parents fell ill. Manser decided to return home immediately. However, the journey was not an easy task for a fugitive. What happened resembled a spy movie. Manser's hair was cut and dyed, and he received blue contact lenses and a forged passport. A total of 22 flights were booked, partly to mislead the police, partly to serve as alternative routes in case of emergency. Bruno Manser, alias Alex Begte, managed to leave Malaysia unharmed.

In Switzerland Manser switched his working methods, but his purpose remained the same: to support the struggle of the Penan people and other native tribes in Borneo. Accompanied by three natives from Borneo, Manser visited 25 cities in 13 countries in a world tour. They gave speeches, met environmental activists, and talked to politicians.

Over the next couple of years, Manser made desperate efforts to bring the plight of indigenous peoples to the world's attention. At the 1991 G7 Summit he climbed a lamppost, chained himself to it, and hung down a banner. At the 1992 Earth Summit in Rio de Janeiro, he paraglided into a crowded stadium during a football match, again with a banner. In 1992 he published *Voices from the Rainforest*, a book describing Penanland, its inhabitants, and their struggle.

In 1993, Manser set up his tent in downtown Bern, and, urging a ban on the import of tropical timber, he started a hunger strike, which he continued for 60 days. Supporters joined him for shorter or longer periods. Some politicians seemed sympathetic, but the ban was not introduced. 'What shocks me is the inconsistency I see in the politicians and economists,' Manser remarked sadly. 'They are personally moved, they see what is going on and in private they fully agree, but they can't support and speak out for it on account of their political status.'[3] However, rainforest destruction received considerable media attention, and people became concerned about the Penan. Moreover, Malaysia officially promised to designate a protected area for the Penan forest dwellers. In 1996 another risky action followed: dressed in a monkey costume, Manser slid down Klein Matterhorn, a 3883 m peak of the Alps, using the cable of the aerial tramway. The spectacular event was accompanied by a press conference about the status of the rainforests and possible ways to protect these important ecosystems.

Between protests and lectures, Bruno Manser returned repeatedly to Borneo to meet his Penan friends. All of his visits were illegal and had to be carried out in secret. Each time he witnessed the rapid shrinkage of the forests. He was also aware that indigenous peoples all over the world were facing the same problems. In 1995 he travelled to Africa, where he spent 2 months among forest-dwelling people. His conclusion: 'Wherever trees are felled, a battlefield is left behind.'[4]

As years went by without significant progress for the rainforest and its inhabitants, Manser got more and more frustrated. He planned to reach a compromise with Malaysian authorities. However, his reconciliatory efforts fell on deaf ears. As a final, desperate attempt, he landed with a motorised paraglider near the palace of Sarawak's chief minister to hand over a toy lamb as a symbol of peace. The chief minister did not show up, Manser was arrested and expelled from the country.

In 2000 he returned to Borneo once again. He crossed the Malaysian border in the jungle and was last seen on May the 25th. Manser arranged a meeting with Penan leaders, but he never arrived, nor did he show any sign of life. Search expeditions were organised by relatives, European colleagues, and Penan friends. They found Manser's last camp, from where his track could be followed for some 600 m in the forest. Then the tracks ended suddenly. Neither Manser's body nor his backpack have been found.

Some thought the disappearance was a trick to attract media attention, but now it is clear that this possibility can be ruled out completely. Manser may have had a fatal accident – the rainforest can be dangerous even for someone who has considerable field experience. It has also been suggested that Manser committed suicide. According to this theory, he found the situation hopeless and wanted to die in the forest. Others maintain that Manser must have been murdered. Whatever the truth is, one thing is certain: those interested in logging were not particularly sad about Manser's disappearance.

---

[3] Quoted in Suter, R. (2015): Rainforest Hero: The Life and Death of Bruno Manser. Bergli Books, Basel, p. 260.

[4] Quoted in Suter, R. (2015): Rainforest Hero: The Life and Death of Bruno Manser. Bergli Books, Basel, p. 233.

In 1950 vast areas of Borneo were covered by pristine rainforests, and the number of nomadic Penan was about 1800. By now most of the forests have been felled and even the last nomadic Penan were forced to settle. Bruno Manser vanished, as did the nomadic Penan. Did Bruno Manser lose the fight? Many battles have been lost, but the war is not yet over. There are rainforests to protect, and there still are some peoples living in the forests. The Bruno Manser Fund continues to fight for the protection of the remaining forests and native tribes.

In one of his talks, Bruno Manser Said, 'How can we influence this negative trend, which without a change of direction will be fatal, especially for coming generations? How can we escape from this vicious circle? I believe that there are some quite simple guidelines and procedures that can help us all. I know from my experience that exactly when we could have made better choices, those choices aren't made because of short-term, so-called economic interests. But the entire economy depends on the end user, for whom it is fairly easy to change a habit or to make a small sacrifice, if you want to call it that. We can forgo and openly boycott various things that are clearly bad – for example products that are connected with human rights abuse or environmental destruction. We can also promote products that respect living creatures and life.'[5]

Recall Manser's words whenever you walk into a shop, and leave the products containing palm oil on the shelf.

Take Action for the Rainforest!

All of us can contribute to rainforest protection by taking some fairly easy steps in our everyday lives. One of the primary causes for logging is wood itself. Do not buy products made of tropical timber. Purchase recycled paper products. A second cause underlying rainforest destruction is agriculture, especially palm oil plantations. Minimise the consumption of foods that contain palm oil. If you buy bananas, cocoa, coffee or similar products from the tropics, pay attention to ecolabels certifying that the farms are being managed sustainably. Large areas of rainforest are turned into pastures to raise cattle, beef being exported to wealthy nations. Alternatively, the land may be used to grow soy bean, which is then exported and fed to cattle, pigs, and chickens all over the world. Thus, reducing meat consumption means protecting the rainforest. Smartphones and other electronic appliances contain rare-earth elements, the mining of which causes deforestation and environmental destruction, especially in the tropics. Buy electronic appliances only if you really need them. Ensure that appliances that you do not need any more are recycled. Tourism is usually a good thing, but it can be harmful if it rests on the exploitation of nature. When you visit the tropics, opt for accommodations and services that minimise their environmental impacts.

---

[5] Quoted in Suter, R. (2015): Rainforest Hero: The Life and Death of Bruno Manser. Bergli Books, Basel, p. 236.

Worth Reading

Hoffman, C. (2018). The last wild men of Borneo: A true story of death and treasure. New York: William Morrow.
Manser, B. (2015). *Voices from the rainforest: Testimonies of a threatened people.* Petaling Jaya: Strategic Information and Research Development Centre.
Suter, R. (2015). *Rainforest hero: The life and death of Bruno Manser.* Basel: Bergli Books.
Straumann, L. (2014). *Money logging: On the trail of the Asian Timber Mafia.* Basel: Bergli Books.

Worth Browsing

bmf.ch
www.survivalinternational.org/tribes/penan

Worth Watching

Blowpipes and Bulldozers (1988)
The Borneo Case (2016)
Bruno Manser (2019)
brunomanser-film.ch/en/

Oil Is Blood – The Fight of Ken Saro-Wiwa

It is a recurring pattern that powerful elite groups benefit from environmental destruction, while marginalised people have to suffer the negative consequences. One of the most extreme cases of this is found in the Niger Delta. And one of the most prominent figures who challenged this pattern was Ken Saro-Wiwa, a businessman and author who became an activist and eventually gave his life for environmental and social justice.

Ken Saro-Wiwa was born in 1941 in the town of Bone, Nigeria. He received an education that was transitional between the traditional and the colonial educations. At school he excelled in both science and sport, but he was especially interested in the English language. Belonging to the Ogoni people, a minority group in multi-ethnic and multi-language Nigeria, Saro-Wiwa felt that English could be a means for uniting all Nigerians. In 1962 he enrolled at the University of Ibadan, where he edited the university's newspapers and was member of the dramatic society and the travelling troupe. After graduating, he began teaching at his alma mater, and later he was lecturer at the University of Nigeria-Nsukka.

In 1966 Nigeria collapsed into anarchy. Following severe ethnic atrocities, the country's eastern region declared independence under the name Republic of Biafra, which resulted in a civil war. When the conflict broke out, Saro-Wiwa was living in Biafran territory. As a supporter of Nigerian unity, he fled to an area controlled by the central Nigerian government. He was appointed civil administrator of Bonny, a city in southern Nigeria. Saro-Wiwa organised food distribution and restarted education. However, his role remains controversial as he refused to investigate cases of military atrocities against civilians. After the civil war ended in 1970, Saro-Wiwa held different posts in the government of the newly created Rivers State, a federal state of Nigeria. He also started various enterprises.

In 1973 Saro-Wiwa left the government and started his literary career. He wrote poems, short stories, TV scripts, plays, and novels, and became well-known both in Nigeria and abroad. His primary goal was to emphasise a common Nigerian consciousness and to contribute to a cultural heritage that could reduce ethnic tensions. One of his most important works, *Sozaboy*, was published in 1985. Written in pidgin

© Springer Nature Switzerland AG 2019
L. Erdős, *Green Heroes*, https://doi.org/10.1007/978-3-030-31806-2_35

English, the internationally renowned anti-war novel is the story of a young boy who becomes a soldier and discovers the absurdity of war. Saro-Wiwa wrote the series *Basi and Company*, which satirised corruption and the desire to get rich quick. Watched by tens of millions of Nigerian citizens, the show became the most successful television programme of the Nigerian Television.

In 1990 Ken Saro-Wiwa abandoned literature and devoted himself to environmental and social causes. Environmental degradation and social injustice were strongly linked in Nigeria. In 1956 oil was discovered in the Niger Delta. Saro-Wiwa's native Ogoni people, as well as other ethnic groups in the Niger Delta, lived from fishing and subsistence farming, thus they depended totally on a healthy ecosystem. After commercial oil drilling started in the Niger Delta in 1958, oil spills became everyday events, making agriculture and fishing impossible, and poisoning the drinking water. In short, oil exploitation resulted in uninhabitable areas. Supported by corrupt politicians, oil companies did nothing to reduce the harmful effects. Cleanups after oil spills were practically non-existent. Gas flaring made the situation even more terrible. As natural gas that accompanies oil reserves is less valuable than oil, it is simply burned, causing air pollution and health problems. To make things worse, oil revenues were redistributed to other regions. Local inhabitants, who had to endure the worsening conditions, received almost nothing in the form of infrastructural or other developments. At the same time, corrupt politicians received huge sums and did not hesitate to create a legal environment that favoured the oil companies. Oil companies were guaranteed land rights almost automatically and had to pay absurdly low compensations for the land owners.

The environmental problems of the Niger Delta attracted Saro-Wiwa's attention as early as 1968. Ecological destruction was a recurring topic in his writings. From 1990, the issue became his primary focus. In that year he set up MOSOP, the Movement for the Survival of the Ogoni People, the aim of which was to organise the peaceful resistance of the Ogoni. He also created several other organisations, including the Ethnic Minority and Indigenous Rights Organization of Africa, which tried to join the forces of different ethnic groups. In 1990 MOSOP created *The Ogoni Bill of Rights*. Signed by all major Ogoni leaders, the document stood for Nigerian unity, but demanded political autonomy for the Ogoni people and emphasised their right to a clean environment. The government ignored the document. Saro-Wiwa's next step was to contact international environmental and human rights organisations. He succeeded in reaching a global audience, but also provoked the disapproval of the Nigerian military regime.

In 1992 Saro-Wiwa published a book titled *Genocide in Nigeria*, in which he argued that the oppression of the Ogoni, combined with the devastation of their environment amounted to genocide. On January the 4th, 1993, some 300,000 people, 60% of the total Ogoni population celebrated Ogoni Day in a mass demonstration organised by Saro-Wiwa. Tension was in the air as no one knew how the military would react to the event. In his speech, Saro-Wiwa declared Shell persona non grata in Nigeria. After the protest was over, the Nigerian security service warned Saro-Wiwa that he would face incarceration if he continued his activism. Saro-Wiwa was arrested several times that year, but he was released, partly because international organisations put pressure on Nigeria.

To secure its operations, Shell hired Nigerian military units, which did not hesitate to use firearms to deter non-violent protesters, injuring and even killing unarmed civilians. This, however, was just the prologue to the unfolding violence against the Ogoni. In 1993 and 1994, the Nigerian military repeatedly attacked Ogoni settlements, killed hundreds of civilians, injured many more, and destroyed the homes of thousands. While Shell's responsibility in hiring military escort is undeniable, the company's role in the later massacres remains unclear.

Once Saro-Wiwa's efforts brought the Ogoni cause to an international stage, some Ogoni leaders, most notably Edward Kobani, started to urge reconciliation with the government and the oil companies, in the hope that the situation would improve. Others, led by Saro-Wiwa, preferred to continue the struggle. The military dictatorship saw the disagreement among Ogoni leaders as a chance to silence the movement. On May the 21st, 1994, Kobani and three other leaders favouring reconciliation were killed. The circumstances of the murders are ambiguous, but many think the events were orchestrated by the military. Ken Saro-Wiwa was arrested and officially accused of having ordered the killings. Fifteen other activists were also arrested. They were tortured and denied access to legal counsel, and held in captivity for months without being charged. It soon became clear that the defendants had no chance of a fair trial. Witnesses were bribed to give false testimonies. Absurd as it may seem, evidence was not required by the tribunal to show that a defendant was guilty. Moreover, the accused were denied the right to appeal the verdict. No one could have any doubt that the sentence was preordained.

Ken Saro-Wiwa was the recipient of the 1994 Right Livelihood Award and the 1995 Goldman Environmental Prize. Being in jail, he was not able to attend the ceremonies; his acceptance speech for the Goldman Prize was smuggled out of prison and read by his son.

On October the 30th, 1995, Ken Saro-Wiwa and eight other activists, Baribor Bera, Saturday Dobee, Nordu Eawo, Daniel Gbooko, Barinem Kiobel, John Kpuinen, Paul Levera, and Felix Nuate were found guilty and sentenced to death by hanging. The Ogoni Nine, as they became known, were executed on November the 10th, despite the heavy international pressure on Nigeria. According to a witness, Saro-Wiwa said before his execution, 'You can only kill the messengers, you cannot kill the message.'[1]

The legacy of Ken Saro-Wiwa lives on. His execution strengthened the resolution of the Ogoni. Also, inspired by Saro-Wiwa's work, other ethnicities in the Niger Delta adopted his tactics and started to fight against the devastation of the environment. Saro-Wiwa redefined corporate accountability. After lawsuits were filed against Shell, the company agreed to pay compensations for the Ogoni. Today even the most powerful corporations have to realise that making profit in an unethical way and causing misery for the local communities can seriously affect their images.

---

[1] Quoted in Mahapatra, D. C. (2004): Dalits in Third Millennium. Sarup & Sons, New Delhi, p. 35.

Unfortunately, however, the situation in the Niger Delta has not improved too much since the 1990s. 'If my father were alive today he would be dismayed that Ogoniland still looks like the devastated region that spurred him to action,'[2] Saro-Wiwa's son noted in 2015. If we consider the negative effects of using fossil fuels, especially oil, global problems have to be taken into account. However, we must not forget about the effects on local ecosystems and societies either. We always have to keep in mind that oil is blood.

Worth Reading

Doron, R., & Falola, T. (2016). *Ken Saro-Wiwa*. Athens: Ohio University Press.
Saro-Wiwa, K. (1992). *Genocide in Nigeria: The Ogoni tragedy*. Port Harcourt: Saros International Publishers.

Worth Watching

Delta Force (1995)
Sweet Crude (2009)
Ken Saro-Wiwa: All for my People (2014)

---

[2] Wiwa, K. (2015): Finally it seems as if Ken Saro-Wiwa, my father, may not have died in vain. The Guardian, November the 10th, 2015. www.theguardian.com/commentisfree/2015/nov/10/ken-saro-wiwa-father-nigeria-ogoniland-oil-pollution

A Passion for Trees – Wangari Maathai and the Green Belt Movement

There are still people who are not totally disconnected from nature, who enjoy the beauty of a blossoming almond tree, who like to eat cherry after climbing the cherry tree, who know how it is to find relief from the boiling sun under a tree, who have experienced the calmness of a snow-covered forest, and who admire magnificent centuries-old trees. These people feel special respect for trees. Trees provide us with fruits, paper, firewood, building material, and a lot of other goods. Trees have an outstanding role in climate regulation, soil protection, and ecosystem health. If we cut trees irresponsibly, we will end up cutting our own throats. But, if used appropriately, trees can remedy multiple environmental problems. This was the formula promoted by Wangari Maathai in Kenya, where tree felling had particularly serious consequences.

Wangari Maathai was born in 1940 in a small Kenyan village. Members of the Kikuyu ethnic group, her parents were peasant farmers. She spent her childhood amid lush vegetation, abundant wildlife, and crystal clear rivers. Moreover, she had her own miniature garden of about 1.5 m², where she cultivated her own plants. She spent a lot of time observing how the crops developed, and took great delight in watching insects and birds that visited the garden. Her enthusiasm towards the soil and the living world never diminished.

Although at that time it was somewhat unusual for girls to go to school, and the costs were substantial compared to the family's income, her mother decided that Wangari should receive formal education. While at secondary school she became interested in science, especially chemistry and biology. It was during her secondary school years when Wangari Maathai first made a big impact. She complained to a priest that many children could not go to school simply because there were no schools in large areas. The priest then spoke to a wealthy British settler, who donated land to the Church, thereby enabling the establishment of a new school.

In 1960 Wangari received a scholarship to study in the United States. In 1963 she celebrated Kenyan independence with fellow students in America. After graduating

© Springer Nature Switzerland AG 2019

L. Erdős, *Green Heroes*, https://doi.org/10.1007/978-3-030-31806-2_36

as a biologist from the University of Pittsburgh in 1966, she returned to Kenya. She was eager to become an active member of Kenyan society. However, upon her arrival she learned that the university job he had been offered had been given to someone else. She was desperate but finally, after a few months, managed to find another job. She earned her PhD in 1971. She was the first woman in Eastern and Central Africa to receive a doctoral degree.

Wangari Maathai started her fight for women's rights when she suggested that female members of the university staff should receive the same benefits as male members – the proposal was strongly resisted by the university.

Her postdoctoral research focused on cattle parasites. During her field work in the Kenyan countryside, she noticed how much the landscape and its inhabitants differed from what she was used to as a child. Vegetation was scarce, trees were being felled, creeks had dried up, rivers were muddy, landslides were frequent, and malnutrition was widespread among both livestock and people. Even her favourite fig tree was cut down, and the nearby spring, where she used to play as a child, disappeared. Problems related to deforestation and erosion also emerged, most notably the permanent shortage of clean drinking water and the lack of firewood for cooking. Maathai participated in the activities of the National Council of Women of Kenya and the Environment Liaison Centre, and with her friends she started to think about the problem and its possible solutions.

Wangari Maathai connected the above environmental and social difficulties with the challenge of unemployment. She thought that the poor could plant trees in and around settlements, and they could gather some income by doing the gardening around the villas of the rich. She set up a tree nursery and planned to sell a fraction of the seedlings to support the project. However, the demand for gardening services and tree seedlings was very low, so the enterprise failed.

Maathai did not give up her plan but started to find alternative routes. She suggested that the National Council of Women of Kenya should start a tree-planting project. The proposal was accepted, and the programme was started on World Environment Day, June the 5th, 1977. Seven trees were planted in a Nairobi park, in honour of seven important figures from Kenyan history. High-ranking politicians attended the ceremony, which received considerable media coverage. The Green Belt Movement was born.

Wangari Maathai soon realised that the long-term success of the project depends on the involvement of local inhabitants who are committed to the movement. The basic idea has remained the same ever since: tree nurseries have been set up all over the country, participants nurture the seedlings, which then are planted to farms, school and church compounds, and degraded areas. Participants receive a small amount of money for each surviving tree seedling. The movement has become very popular, with participants working hard to turn Kenyan landscapes green by planting trees – primarily native species.

Native Versus Non-native Trees
Unfortunately, non-native trees are usually preferred to native ones these days. Some want to have exotic plants in their gardens, while fast-growing non-native trees often enjoy preference in forestry. But it is important to keep in mind that non-natives have a lot of disadvantages. Native trees attract local wildlife, as numerous species in the region have adapted to use the resources provided by these trees. Native trees serve as centres of intricate ecological networks, and may support hundreds of different living beings, from birds to bacteria. In contrast, newcomers do not fit the local ecosystems as they have a completely different evolutionary history. This is one reason why non-native plantations are almost empty, while native forests are full of life. If you prefer a living garden, opt for native trees!

While Wangari Maathai gained national and international acclaim through the Green Belt Movement, she was considered a troublemaker by the Kenyan government. What made her especially suspicious was her speaking up for women's rights and democracy. In 1982 she wanted to run for a seat in Parliament but was disqualified. She also lost her university job. Nonetheless, these first conflicts also had a positive effect, as from that point on Wangari could devote all her time to the Green Belt Movement. Funding from abroad enabled the expansion of the program. Now they could afford to employ local people to take care of the tree nurseries. The Green Belt Movement also started workshops where people discussed environmental and social problems, and were encouraged to do their best to improve the situation. In the 1980s the movement took root in other African countries.

Maathai's confrontation with the government deepened in 1989. When the plans to erect a skyscraper in Nairobi's Uhuru Park came to light, Maathai decided to take action. Uhuru Park is to Nairobi what Hyde Park is to London and Central Park to New York. It turned out that the skyscraper would have consumed a large proportion of the park. In addition, two historic buildings were doomed to demolition. The plan was also absurd because infrastructural issues such as traffic and parking were not considered. Maathai wrote to different authorities, and informed the media about the adverse effects of the planned building. As a response, government officials and other politicians ridiculed her. The Green Belt Movement, which rented an office in a government building, was thrown out and the staff had no choice but to move to Maathai's flat. After a while the struggle was not only about the park. The park itself symbolised democracy, while the skyscraper became the symbol of the authoritarian regime. Eventually, construction plans were cancelled and the park was saved.

In 1992, Wangari Maathai, by then an eminent member of the Kenyan pro-democracy movement, was arrested. Later, the charges were dropped after eight US senators, including Al Gore, had warned the Kenyan president that by harassing democracy advocates he would risk damaging the relations between Kenya and the

US. Hardly had Maathai been released from jail when she organised a demonstration that demanded the release of political prisoners. The relatives of the prisoners and other supporters of the cause gathered in Uhuru Park. They remained peaceful, but the police attacked them violently, using tear gas and beating the protesters with truncheons. Wangari Maathai and some other participants suffered bad injuries. The demonstration was resumed in a nearby church and lasted until the political prisoners were freed. Today the section of the Uhuru Park where the demonstration was crushed is called Freedom Corner.

During the following years Maathai protested whenever public land was given to private investors for luxury developments, a practice known as 'land-grabbing'. Some of her campaigns were successful, others failed. The most famous incident happened in 1998. Maathai was informed that Karura Forest, a public land in the proximity of Nairobi, was slated for development by the allies of the government. The area had remarkable biodiversity and was also important for the inhabitants of the nearby city. The Green Belt Movement immediately started a campaign. First they appealed to the authorities, then they raised public awareness. As a next step they went into the area and started to plant trees. Their efforts were supported by students, opposition politicians, environmental and human rights organisations, and the Church. During one of the tree-planting rallies, the demonstrators were attacked by private security guards, and the police did not intervene to protect them. Maathai was injured once again, as were some other participants. Soon riots broke out in Nairobi, and the government decided to declare the forest protected to prevent the escalation of violence. Today Karura Forest is a woodland of high natural value that also serves as a recreational area. Local inhabitants are employed as tour guides and everyone is proud of this small wilderness next to the Kenyan capital.

After Kenya's transition to democracy, Wangari Maathai was elected Member of Parliament in 2002, receiving 98% of the votes in her constituency. From 2003 to 2007 she served as Assistant Minister for Environment and Natural Resources.

Wangari Maathai received countless prestigious awards for her work, including the Right Livelihood Award (in 1984), the Environment Medal of the Better World Society (in 1986), the UN Global 500 Award (in 1987), and the Goldman Environmental Prize (in 1991). In 2004 she was awarded the Nobel Peace Prize 'for her contribution to sustainable development, democracy and peace.'[1] She was underway to her constituency when she received a call that informed her about the decision of the Nobel Committee. She interrupted the journey and celebrated the good news by planting a tree.

Wangari Maathai died in 2011. Her impact is immense. The Green Belt Movement has established thousands of tree nurseries, mobilised hundreds of thousands of participants, and planted tens of millions of trees. Moreover, Wangari Maathai showed that it is possible to reduce poverty and combat environmental destruction simultaneously by community-based programmes such as her tree-planting project.

---

[1] www.nobelprize.org/prizes/peace/2004/maathai/facts/

She also showed how closely human rights violations and environmental destruction are linked.

'The future of the planet concerns all of us, and all of us should do what we can to protect it,'[2] Wangari Maathai wrote in her autobiography. Wangari Maathai, alias 'Mama Trees' did exactly that.

Worth Reading

Maathai, W. (2004). *The Green Belt Movement: Sharing the approach and the experience.* New York: Lantern Books.

Maathai, W. (2007). *Unbowed: A Memoir.* New York: Anchor Books.

Maathai, W. (2010a). *Replenishing the Earth: Spiritual values for healing ourselves and the world.* New York: Doubleday Religion.

Maathai, W. (2010b). *The challenge for Africa.* London: Arrow Books.

Worth Browsing

www.greenbeltmovement.org

Worth Watching

Wangari Maathai: The Eco-Warrior with a Smile (2012)

---

[2] Maathai, W. (2007): Unbowed: A Memoir. Anchor Books, New York, p. 138.

Vandana Shiva – Defending Traditional Agriculture

In the 1960s and 1970s the agriculture of many developing countries was character-ised by a process known as green revolution. Despite its name, green revolution had nothing to do with environmental protection. On the contrary, the indiscriminate introduction of western technologies resulted in massive pollution and also had harmful social consequences.

The destructive forces of the green revolution have been particularly strong in India. Indian subsistence farmers used to grow numerous crops, mainly local variet-ies, in diverse agro-ecosystems. During the green revolution, these were replaced by monocultures of a few high-yielding varieties, which needed mechanisation, chemi-cal fertilisers and pesticides, usually also irrigation. The resulting soil and water pollution was made worse by the fact that farmers did not know anything about correct pesticide dosage or safety precaution methods. Indian women would gather a large number of weeds, which they used as vegetables and fodder. The eradication of weeds through heavy chemicalisation made this practice impossible. While tradi-tional agriculture needs almost no investment, industrial agriculture is quite capital-intensive: farmers have to buy tractors and other equipment, fertilisers and pesticides. Farmers ran up huge debts, and the situation of many became totally hopeless due to the high interest rates. Also, monocultures were more prone to weather and mar-ket fluctuations than diverse traditional farming systems. The independence of the farmers disappeared alongside the independence of the regions: the seeds of high-yielding varieties, machinery and pesticides had to be imported, and irrigation depended on how much water was left in the river by upstream usage.

Today, even the advocates of the green revolution admit that it had serious nega-tive impacts, but they maintain that the overall balance was positive, as it helped feed rapidly increasing populations. Critics argue that the green revolution did more harm than good, and farmers' situation is worse now than it was prior to the green revolution. According to this latter logic, it would have been better to return to tra-ditional methods that are adapted to local conditions and are able to produce food without hazarding ecological and social integrity. Vandana Shiva is probably the best known proponent of this view.

© Springer Nature Switzerland AG 2019
L. Erdős, *Green Heroes*, https://doi.org/10.1007/978-3-030-31806-2_37

Vandana Shiva was born in Dehradun, India, in 1952. She developed an interest in physics at the age of nine and graduated as a physicist. Later she studied philosophy of science in Canada, earning a PhD at the University of Western Ontario in 1978. She first met environmentalism when, as a student, she joined the Chipko movement. This was a campaign of rural Indian women who prevented logging by embracing trees.

In 1982 Shiva set up the Research Foundation for Science, Technology, and Natural Resource Policy, the mission of which is to put scientific results at the service of sustainability and social justice. Initially the Foundation operated in the cowshed of Shiva's mother. The organisation was later renamed Research Foundation for Science, Technology and Ecology.

In 1991 Shiva launched Navdanya ('nine seeds' in Hindi), a movement that intends to protect traditional agriculture and the biodiversity of local crop varieties. Navdanya has created more than hundred seed banks, and collected and saved thousands of local plant varieties, many of which are able to withstand drought or other extreme environmental conditions. Seed banks are set up by Navdanya but are managed by local communities. From the seed banks, seeds are distributed among farmers. Also, farmers are encouraged to exchange their own seeds and associated knowledge. Navdanya informs farmers about the risks of chemical pesticides and supports organic agriculture and fair trade, while it also promotes healthy nutrition and the idea of eating local food. The organisation operates its own farm, which serves as a training centre.

Vandana Shiva is a prolific author, having published more than 20 books. In *The Violence of the Green Revolution*, one of her most influential works, Shiva summarised why the green revolution created more problems than it solved. According to Shiva, Indian agriculture functioned well until it was disrupted during the era of colonialism. When the nation became independent, there were some timid attempts to restore traditional agriculture. However, these were overwhelmed by a 'modernisation' that was totally alien to Indian culture and history. Industrialised agriculture was imported from the western world, with all of its problems and few of its benefits. Shiva emphasises that high-yielding varieties perform better than traditional crops only if there is a huge input of fertilisers, pesticides, irrigation, and fossil fuels. The cost for a slight increase in yield includes biodiversity loss, water shortage, pollution, farmers' indebtedness, social conflicts, and serious health effects. The winners of the intensification were western chemical companies, which gained new markets for their products.

Although the consequences of the green revolution proved disastrous, the idea may have been well intentioned. Genetic modification, according to Vandana Shiva, is fundamentally different in that it has one purpose only: to increase the profits of a few multinational corporations. No one denies that, in theory, genetic modification may result in higher yields in smaller areas, which may be good for relieving famine and preserving nature. However, up to now no genetically modified organism (GMO) has achieved these goals, nor can it be expected that we will ever have such a GMO. As Lester Brown, the founder of the Worldwatch Institute put it '... no genetically modified crops have led to dramatically higher yields ... Nor do they seem likely to do so, simply because conventional plant-breeding techniques have already tapped most of the potential for raising crop yields.'[1] In addition, pesticide-resistant GMOs cause massive pollution, as high pesticide dosages can be used to kill everything except for the GMO. A further problem associated with GMOs is that they can easily contaminate the fields of farmers who oppose this technology. One of the most shocking examples is the case of Percy Schmeiser, a Canadian farmer who grew non-GMO canola. In 1997 his farm was contaminated by Monsanto's GMO canola. Schmeiser was sued by Monsanto for patent infringement, and was found guilty.

Vandana Shiva emphasises that working with nature would be able to alleviate famine and poverty without the risks of GMOs. Traditional agriculture, diverse farming systems, and local varieties provide more food of better quality than GMOs. For example, instead of using a GM rice that produces the precursor of vitamin A, there is a much better solution to fight vitamin A deficiency, namely the utilisation of traditional crops with a naturally high concentration of the same precursor. While the health effects of GMOs are debated, their ecological effects are mostly negative. Also, even if some advantages of GMOs are acknowledged, genetic engineering should by no means be regarded as a wonder method that can solve humanity's nutritional and environmental problems.

In many countries, creativity and innovation are protected by intellectual property rights. However, neither the information encoded in natural organisms' DNA sequences nor the traditional knowledge of indigenous tribes and rural farmers are protected under these laws. Thus, the wisdom that nature, tribal peoples, and traditional communities have accumulated through countless generations can be expropriated by western corporations. These corporations develop medicines or other products, based on biodiversity and its traditional usage by local inhabitants. Subsequently, they protect 'their' inventions by patents, and sell the product, making huge profits. Local inhabitants receive nothing, although it was they who originally recognised the usefulness of the organism and developed methods to utilise it. Vandana Shiva calls this practice biopiracy and regards it as a new form of colonialism. She calls for legal changes to end this form of exploitation. Shiva points out that biopiracy is not only unfair, but also destructive. Traditional societies use biodiversity in a smart way, ensuring the continued existence of diverse ecosystems. In contrast, biopiracy tends to use biodiversity as a source for introducing monocultures, which destroy diversity.

In India as well as many other countries, women play a major role in agriculture. Also, women are responsible for providing the family with food and drinking water. Consequently, it is women who first face environmental degradation in the form of soil contamination, erosion, water scarcity, and the lack of fruits or firewood. Shiva explores this topic in her book *Staying Alive: Women, Ecology, and Development*. As one of the most influential ecofeminist writers of our time, Shiva shows that the oppression of women and the destruction of nature have the same mental roots. As

---

[1] Quoted in Nader, R. (2014): Food Science: What's the Harm? Huffington Post, September the 15th, 2014. www.huffpost.com/entry/food-science-whats-the-ha_b_5824036

a possible solution to the pressing agro-environmental problems she suggests that nature should be treated as a partner rather than an enemy. Led by multinational corporations, agribusiness wages war on nature, deploying huge amounts of herbicides, fungicides, and insecticides. A more friendly farming, usually conducted by rural women, knows how to preserve biodiversity and produce healthy and nutritious food. Other living beings are treated with respect. As emphasised in Shiva's book *Stolen Harvest*, 'cattle and earthworms are our partners in food production.'[2]

In her book *Monocultures of the Mind* Shiva explores how and why diverse habitats and farming systems are replaced by monocultures, and analyses the risks of producing large-scale homogeneity and uniformity throughout the world. It is a mistake to treat crop yield and timber as the only measures of economic value. In addition to the main crop yield, a traditional farm also provides complementary nutritional sources (in the form of weeds used as vegetables), and fodder, while it preserves soil fertility. Similarly, forests provide not only timber, but also firewood, fruits, and countless ecosystem services such as soil protection. While monocultures maximise one kind of output, they minimise all the other outputs. If the value of the total output is taken into account, diverse forests prove more valuable than artificial tree plantations, and traditional farms are better than crop monocultures. The claim that only industrial agriculture can feed the world is usually repeated like a mantra. However, in her books *Soil Not Oil* and *Who Really Feeds the World?* Shiva shows that small-scale, ecological agriculture is able to produce the same amount of food on a unit area.

The Importance of Biodiversity for Agriculture

In most of the world, the majority of arable lands are dominated by only three crops: wheat, maize, and rice. While these plants have several varieties, only a handful of them are used for large-scale cultivation. One may think that a few varieties of a few species are enough for human needs, but reality is much more complicated.

Preserving the diversity of plants is a necessary part of food security. Plant diversity is essential at three levels. First, we have to protect wild plants that are not currently used as food, but which may become important in the future, especially as environmental conditions or human needs change. By destroying natural habitats we risk losing a large number of potentially useful plants. Second, it is essential to protect the wild ancestors and relatives of cultivated species. The gene pools of these species can be used to breed new varieties that are resistant to certain pests or are able to tolerate harsh environments. Third, the huge diversity of cultivated varieties and ancient strains needs to be conserved. Not too long ago, every region had its own fruit and crop varieties that were adapted to local conditions. Every variety had its own typical form, colour, and savour. Some of these valuable varieties survive in small-scale farms and seed banks, and can occasionally be found in local markets.

---

[2] Shiva, V. (2016): Stolen Harvest: The Hijacking of the Global Food Supply. University Press of Kentucky, Lexington, p. 12.

The poor are particularly vulnerable to the effects of environmental degradation. Therefore, Vandana Shiva thinks environmental and social problems should not be treated separately. A sustainable society is a just society, and vice versa. 'Equality and sustainability are two sides of the same coin,'[3] she insists.

Vandana Shiva is an icon of ecofeminism and one of the most respected environmentalists of the twenty-first century, an influential thinker and an on-the-ground activist at the same time. In some respects, her activity is akin to that of Mahatma Gandhi, who is her role model and source of inspiration. Vandana Shiva has received countless awards, including the Right Livelihood Award and the Global 500 Award. She has appeared in dozens of documentaries. David Brower, the outstanding American environmentalist, founder of Friends of the Earth once said that if there was a world president, he would vote for Vandana Shiva.

Although she is a constant target of those who support industrialised agriculture and genetic manipulation, Shiva continues to work tirelessly. 'I live by my conscience. And that's why I've managed to keep going,' she says.

Worth Reading

Shiva, V. (1993). *Monocultures of the mind: Perspectives on biodiversity and biotechnology.* London: Zed Books.

Shiva, V. (2000). *Stolen harvest: The hijacking of the global food supply.* Cambridge: South End Press.

Shiva, V. (2013). *Making peace with the Earth.* London: Pluto Press.

Shiva, V. (2015). *Soil not oil: Environmental justice in an age of climate crisis.* Berkeley: North Atlantic Books.

Shiva, V. (2016a). *The violence of the Green Revolution: Third World agriculture, ecology, and politics.* Kentucky: University Press of Kentucky.

Shiva, V. (2016b). *Biopiracy: The plunder of nature and knowledge.* Berkeley: North Atlantic Books.

Shiva, V. (2016c). *Staying alive: Women, ecology, and development.* Berkeley: North Atlantic Books.

Shiva, V. (2016d). *Who really feeds the world? The failures of agribusiness and the promise of agroecology.* Berkeley: North Atlantic Books.

Worth Browsing

www.navdanya.org

---

[3] Vandana Shiva at the Bonn Conference for Global Transformation, 2015. www.youtube.com/watch?v=ecBIeKd5KII

Worth Watching

Seeds of Life (2003)
The World According to Monsanto (2008)
Mad Cow Sacred Cow (2009)
Seeds of Freedom (2012)
Seeds of Death: Unveiling the Lies of GMO's (2012)
GMO OMG (2013)
Revolution Food (2015)
Seed: The Untold Story (2016)

Tree Huggers and Hunger Strikers – Environmental Leadership in India

The phrase 'tree hugger' is often used to ridicule environmentalists. But if we trace the origins of tree huggers, it will be evident that the expression is not offensive at all.

Until relatively recent times, the Himalaya ranges of northern India were quite hard to access and, consequently, relatively isolated and more or less intact. The 1960s, however, brought a rapidly growing interest in the region's natural resources, including timber and various minerals. The intensive deforestation that followed caused landslides, soil erosion, the drying up of water sources, and a lack of firewood, fodder, mushrooms, forest fruits, and medicinal herbs. Both the ecological balance of the area and the traditional lifestyle of the mountain people were threatened. Large corporations made sizeable profits, whereas the life of locals was devastated. The final straw was when, in 1972, the inhabitants of a small settlement were denied permission to fell a few trees for making tools, while a company was allowed to fell much more trees in the same area. Desperate and incensed, the villagers decided to protect the trees by embracing them. Their non-violent demonstration was successful, as the company's right to cut the trees was cancelled.

The approach spread to other villages and quickly gained international acclaim. The new grass-roots movement became known as Chipko, meaning 'to hug' or 'to adhere to' in Hindi. The Chipko movement proved highly successful: not only was it able to save trees and forests at several locations, but it also influenced the government's forestry policy, resulting in regionally decreasing deforestation.

The man who organised the initial protest and may therefore be regarded as Chipko's initiator is Chandi Prasad Bhatt. But the backbone of the movement was formed by women. The most famous of them, Gaura Devi, became an active member of Chipko spontaneously. She was living in Reni, a small village in the Indian Himalayas close to the Tibetan border. On March the 25th, 1974, all men of the village were away in a town to collect compensation for their land used by the army. Precisely on that day, loggers arrived near Reni to cut down trees, as permitted by the Forest Department. Luckily, a young girl spotted the lumbermen and immediately hurried to Gaura Devi, leader of the local women's organisation. Gaura Devi did not hesitate to take action: she recruited women and children from the village

© Springer Nature Switzerland AG 2019
L. Erdős, *Green Heroes*, https://doi.org/10.1007/978-3-030-31806-2_38

and led them to the forest. Upon meeting the loggers, the women explained them how important the forest was for their livelihood. Devi and the other women refused to leave the area and protected the trees with their bodies. When one of the workers, probably drunk, aimed a gun at her, she said: 'This forest nurtures us like a mother; you will only be able to use your axes on it if you shoot me first.'[1] After a standoff lasting for a couple of days, the loggers withdrew.

The success of Chipko, to a large degree, depended on spreading its message from one village to another. Sunderlal Bahuguna played a major role in that process.

Bahuguna was born in 1927. Aged 13 he met a disciple of Gandhi and joined the independence movement. Later he started to work as a journalist and also became involved in politics. He abandoned his party when he married, as his wife insisted that he leave the political arena and they settle in a remote village. Bahuguna became an environmentalist after he realised how fatally deforestation affects rural inhabitants and the environment. When the Chipko movement unfolded, Bahuguna took the idea from village to village in the Indian countryside. He also participated in demonstrations. On one occasion, when he was sleeping in a forest, guarding the trees, his tent was set on fire by someone who did not agree with the goals of the Chipko movement.

In 1981 Bahuguna undertook a march through the Himalayas. He covered a distance of 4870 km, walking from Kashmir to Kohima. On his way he met a lot of people, whom he told about the importance of protecting forests. Bahuguna's resolution has not diminished and his campaign for Himalayan forests continues to this day.

Sunderlal Bahuguna spearheaded the movement opposing the construction of a huge dam on the Bhagirathi River, near Tehri. Though colossal dams were seen as symbols of progress by the government, environmentalists pointed out that these megalomaniac projects had numerous flaws. The electric power generated by the dams was needed mainly by industry, whereas the water was intended for the export-oriented sections of agriculture and for cities with already high water consumption. Thus, the local rural population gets none of the potential benefits of the dams. Meanwhile it is precisely these rural inhabitants who have to suffer all the negative consequences. Several settlements and huge areas of ancestral lands, including forests and small-scale agricultural plots, were destined to be flooded. About 100,000 people were planned to be evicted. In addition, the river is considered sacred by the local population; to build a dam on it is a sin. Also, the dam and the reservoir have terrible ecological consequences. Moreover, as the area is earthquake-prone, many fear the dam presents a hazard for people living downstream. Of course, environmentalists accept that the country needs electricity and water sources, but they insist that there would be more cost-effective and ecologically and socially less harmful solutions.

---

[1] Desmond, K. (2008): Planet Savers: 301 Extraordinary Environmentalists. Greenleaf Publishing, Sheffield, p. 197.

Bahuguna started a series of hunger strikes on the river bank (he prefers to call these fasts, because he insists they have a spiritual dimension). Led by Bahuguna, a group of activists blocked the road leading to the construction site, whereupon they were arrested. As Bahuguna was a renowned and widely known environmentalist, his protests attracted considerable attention, and there were promises from the government that the Tehri dam project would be reviewed by an independent body. Unfortunately, the promises were not kept. While the demonstrations were able to slow down the construction, they could not bring it to a halt. The dam was built; Bahuguna himself was evicted and his home is now under water.

However, Bahuguna did not give up his hope of saving the Himalayas from reckless exploitation. His dream is to see the region being developed in a way that respects its natural beauty and the rural communities. He advocates a new strategy for the Himalayas: a wise use of resources based on local traditions, instead of mistaken mega-projects, intensive deforestation, mining, and monocultures.

Sunderlal Bahuguna is an archetypal Gandhian activist: he is non-violent, humble, perseverant, cheerful, and wise. Bahuguna keeps saying that today we have plenty of knowledge but not much wisdom, a scarcity of love, and only a limited willingness to take action. He thinks there is a need for balance among head, heart, and hands. 'Use your head for creative thinking, your heart for compassion, and your hands for honest action,'[2] India's green guru encourages environmentalists and all right-minded people.

The Tehri dam is rightly considered infinitely megalomaniac, but how should one describe the Narmada Valley Project, which includes the construction of 30 large and over 3000 smaller dams on the Narmada River and its tributaries? Needless to say, the environmental and social impacts of the project are immense, including the displacement of more than one million people and the submergence of vast areas of forests and farms. Of the activists opposing the project and advocating ecologically and socially sound alternatives, Medha Patkar is the best known.

Medha Patkar was born in 1954 in Bombay (today Mumbai). As a PhD student she studied the effects of the Narmada Valley Project. While proponents of the dams emphasised the project's potential benefits (hydroelectric power, water for irrigation, and drinking water), environmentalists pointed out the extremely high financial costs, the possibility of corruption, the destruction of wildlife, the lack of consultation with the local inhabitants, the loss of forests and agricultural areas, the eviction of a huge population of marginalised people without fair compensation, and the short lifespan of the dams due to the high rate of sedimentation. Environmentalists claim that the benefits were systematically and seriously overestimated, while the social and environmental costs of the project were consistently underrated. Also, it turned out that there was no environmental impact assessment for the project.

Medha Patkar thought fighting against the dams was more important than doing research. Therefore, she quit her doctoral studies and started to mobilise people. She

---

[2] Sunderlal Bahuguna in the 2005 documentary Appiko: To Embrace.

met a huge number of local residents, listened to their complaints and presented their concerns to the authorities. In 1986 she organised a demonstration in the form of a 36-day march. Despite the violent attack of the police, the march was a great success, as it managed to raise considerable awareness about the issue.

There were many people who opposed the dam: local inhabitants who faced losing their homes and lands, fisherfolk, human rights activists, environmentalists, and scientists. To unite them, Patkar founded the Narmada Bachao Andolan (Save the Narmada Movement), which organised roadblocks and protests, and started lobbying the World Bank not to back the project. Medha Patkar and some other activists even went on hunger strikes. Though the government promised to carry out a comprehensive review, nothing happened for a long time. As the opposition became more and more visible, the World Bank decided to re-evaluate its position and, in 1993, withdrew from the project. But the state government pushed on and decided to fund the project itself.

In 1994 the Narmada Bachao Andolan appealed to the Supreme Court of India, challenging the construction of the Sardar Sarovar Dam, the largest dam in the series and the flagship of the Narmada Valley Project. The construction works were suspended, but after years of legal battle, the Supreme Court decided that the dam could be built. The Sardar Sarovar Dam was inaugurated in 2017. It is one of the largest dams in the world.

Medha Patkar's fight continues: her aim is to achieve that all displaced people receive just compensation. She is also involved in numerous other projects in the fields of social justice, democratic rights, and sustainable development. For her unflagging work Patkar has received numerous awards, including the Right Livelihood Award, the Goldman Environmental Prize, and the Mother Teresa Memorial International Award.

According to Patkar, the subjugation of nature is not development but destruction. 'If the vast majority of our population is to be fed and clothed, then a balanced vision with our own priorities in place of the Western models is a must. There is no other way but to redefine "modernity" and the goals of development, to widen it to a sustainable, just society based on harmonious, non-exploitative relationships between human beings and between people and nature,'[3] she says.

Rivers are not only threatened by dams, but also by increasing pollution. The situation is particularly acute in India. The case of the Ganges River has received worldwide attention. Though the river has a diverse wildlife and enormous religious and cultural importance, massive pollution and dams threaten its ecological integrity and sacredness. Fortunately enough, the Ganges has always had its protectors. Guru das Agrawal was one of the most notable of them.

Guru das Agrawal was born in 1932. He graduated from the University of Roorkee and earned his PhD in environmental engineering at the University of

---

[3] Quoted in Mukherjee, T. (2017): How Fares the Well? A Study of the Interstices of the Welfare State: Bharati Sarabhai's The Well of the People (1943), Mahasweta Devi's Jal/Water (1976), and Vinodini's Daaham/Thirst (2005). Journal of Commonwealth Literature July–September 2017: 1–10.

California, Berkeley. He worked as the head of the Department of Civil and Environmental Engineering at the Indian Institute of Technology, Kanpur, and served as technical director at a company specialising in environmental impact assessments.

As an environmental activist, Agrawal was known primarily for his hunger strikes. In 2009, protesting against the hydroelectric dams planned on the Bhagirathi River, one of the source rivers of Ganges, Agrawal went on hunger strike. After he spent more than a month without food, the plans to build the dams were cancelled.

The government of India started to carry out a plan to save and restore the Ganges River. Despite the large budget, the action proved largely ineffective due to corruption and poor implementation. Disillusioned with the weak results, Agrawal wrote to the prime minister in 2018, demanding immediate action in favour of the Ganges. As he received no response, he started a hunger strike. He died of starvation on the 111th day of the fast.

Worth Reading

Weber, T. (1989). *Hugging the trees: The story of the Chipko movement*. Harmondsworth: Penguin Books.
James, G. A. (2013). *Ecology is permanent economy: The activism and environmental philosophy of Sunderlal Bahuguna*. Albany: State University of New York Press.

Worth Watching

Water Wars (2009)
Appiko: To Embrace (2005)
Drowned Out (2004)
DAM/AGE (2002)

The Sacrifice of Berta Cáceres

Each year up to 200 environmentalists are killed because of their activism, mostly in Latin America, Southeast Asia, and Africa. The primary cause is greed, corruption, and unstable political circumstances in countries where both the environmental legislation and law enforcement are poor. However, the ultimate cause is the world's hunger for energy, minerals, fossil fuels, palm oil, ivory, and similar products. Indigenous leaders, activists, and rangers who want to protect their ancestral lands, environmental integrity or natural values, risk their lives. But nowhere else are they at greater risk than in Honduras, the deadliest country for environmentalists. Tragically, most assassinations do not cause much stir in the world's media. However, the killing of Berta Cáceres prompted an international outcry and called for increased protection of indigenous rights.

Berta Cáceres was born in 1971 in La Esperanza, Honduras. She belonged to the Lenca group, an indigenous people of Central America. Social sensitivity was instilled in Berta during her childhood: her mother cared for underprivileged rural people who had poor access to education and healthcare, and also supported Salvadorean refugees during the civil war in El Salvador.

From the 1980s and especially the 1990s, the lands of Honduran indigenous communities and small-scale farmers have been under increasing pressure from illegal logging, hydroelectric dams, mining operations, palm oil plantations, and luxury tourist resorts. Although native groups have been living in the area for hundreds or thousands of years, they often lack written documents, and, consequently, their land rights are not officially recognised. They were deprived their territories without consultation and compensation. As native groups tried to resist, the powerful elite groups interested in the mega-projects reacted violently.

After the military coup in 2009 the situation in Honduras was rapidly deteriorating: corruption, drug trafficking, sexual crimes, and murders were on the rise, as was poverty. The overwhelming majority of violent crimes went unpunished. The use of indigenous land for business interests was booming. Indigenous communities were forcefully evicted from their ancestral lands. Environmental activists were in

© Springer Nature Switzerland AG 2019
L. Erdős, *Green Heroes*, https://doi.org/10.1007/978-3-030-31806-2_39

great danger: dozens of them were brutally murdered. At the same time, authorities started to treat environmentalists as troublemakers at best, criminals or terrorists at worst.

As a student, Berta co-founded the Council of Popular and Indigenous Organizations of Honduras (COPINH) in 1993. From that time she devoted most of her energy to help indigenous communities in protecting their lands, livelihoods, and environments. Her first campaign managed to stop an illegal logging operation in the west of Honduras. In July 1994, she organised the Indigenous and Black Pilgrimage for Life, for Justice and for Liberty, which was attended by tens of thousands of participants. It was a historic event because it united different marginalised groups and drew attention to their plight. During the government of Carlos Roberto Reyna, COPINH achieved some results in improving indigenous communities' access to healthcare and education, as well as in the recognition of indigenous rights. Due to several protests led by Berta in 2006 and 2007, plans to build a new dam on the Lempa River between Honduras and El Salvador were dropped.

As Berta Cáceres succeeded in stopping several dam, logging, and mining projects, she became a respected environmentalist both in Honduras and internationally. But her ultimate project was the fight against a cascade of dams on the Gualcarque River.

The story began when a Lenca community informed Cáceres that heavy machinery and construction equipment had been shipped to their land. Cáceres started to investigate and found out that a series of four hydroelectric dams were planned on the Gualcarque River. Not only is the river a source of water and fish for a large number of Lenca families, but it is of enormous cultural significance as well, for it is regarded as sacred by the Lenca people. In addition, the construction works also threaten large tracts of forest and small-scale agricultural lands in the proximity.

Berta handed in official complaints, and started to organise protests. She brought the struggle to an international stage, highlighting the social and environmental disaster the project would cause.

In the spring of 2013 Berta Cáceres started a peaceful blockade, which, for more than a year, was able to prevent the hydroelectric firm from accessing the site and beginning the construction works. In July 2013 the military opened fire on the unarmed protesters, killing local indigenous leader Tomás García Dominguez and injuring others. Eventually, the human rights violations backfired, as both the main financing partner and the Chinese building company withdrew from the project.

The success, however, came at a high personal cost for Berta and the other activists. The company and its allies, including the military and the political elite, were determined to push forward with the dam, even if this meant the complete elimination of the opposition. Berta was offered money to stop her activities but refused. From 2013 she was increasingly being intimidated by the security employees of the company, suffered wrongful arrests based on fabricated charges, and received dozens of death threats. The risk to her life was real: according to international estimates, about one hundred environmental campaigners had been killed in Honduras during the previous few years alone. The 2014 analysis of Global Witness concluded that Honduras was 'the most dangerous country to be an environmental defender.'[1] Berta herself was well aware of the danger. 'When they want to kill me, they will do it,'[2] she said in an interview. But she continued the struggle, noting that 'They are afraid of us because we are not afraid of them.'[3]

In 2014 Berta met Pope Francis at the World Meeting of Popular Movements, initiated by the Holy Father to bring together grassroots organisers. In 2015 Berta received the Goldman Environmental Prize, one of the most prestigious awards for green activists. 'Our Mother Earth – militarised, fenced-in, poisoned, a place where basic rights are systematically violated – demands that we take action,'[4] she said in her acceptance speech. Some hoped that international recognition would protect Berta's life. But these hopes proved unfounded. As the works on the dam started again with new investors in 2015, and anti-dam protests also resumed, the company feared Berta's work would cause further delay. She was marked for death.

On March the 2nd, 2016, just before midnight, gunmen broke into the home of Berta Cáceres and shot her dead. Gustavo Castro Soto, a Mexican environmentalist who was staying with Berta to discuss environmental topics was also shot but survived the attack.

The assassination of Honduras' best-known environmentalist – and especially the failure of the police to protect her – provoked a mass demonstration of university students in Tegucigalpa, the capital of Honduras. Berta's funeral was attended by thousands of people and became a demonstration against violence.

Initially the authorities claimed the attack had been a robbery. Later they wanted to blame COPINH for the murder but the theory was not supported by any evidence. Subsequent investigations showed that the company behind the dam might have been involved in the crime. In 2016 and 2017 eight persons were arrested in relation to the murder, seven of whom were found guilty. Some of them were connected to the hydroelectric company, others to the Honduran military. The suspected intellectual author of the murder, the president of the hydroelectric company, was arrested in 2018 and faces trial separately. Unfortunately, the role of high-ranking politicians and the military has not been investigated, and the real network responsible for the murder of Berta Cáceres and other activists has remained untouched. Multinational corporations and development banks supporting the project were also part of the pattern that caused Berta's death, as they, for a long time, ignored serious complaints about human rights and environmental issues. In addition, it would be

---

[1] www.globalwitness.org/en/campaigns/environmental-activists/how-many-more/

[2] Lakhani, N. (2013): Honduras dam project shadowed by violence. Al Jazeera, December the 24th, 2013. www.aljazeera.com/indepth/features/2013/12/honduras-dam-project-shadowed-violence-201312211490337166.html

[3] Quoted in Petermann, A. (2018): Earth Minute: Remembering Berta Caceres. globaljusticeecology.org/earth-minute-remembering-berta-caceres-judi-bari-on-international-womens-day/

[4] Quoted in Schachet, C. (2016): Berta Cáceres Vive! Cultural Survival Quarterly Magazine, June 2016. www.culturalsurvival.org/publications/cultural-survival-quarterly/berta-caceres-vive. The video of the acceptance speech is available at: www.goldmanprize.org/recipient/berta-caceres/

hard to deny the responsibility of foreign governments that failed to condemn Honduran authorities despite the ongoing terrorisation of environmental and human rights activists.

Thanks to courageous activists, COPINH is operating, in spite of the continuing harassments, death threats, and murders. So far, the organisation has managed to stop some 50 major logging and 10 dam projects, and they achieved the designation of some protected areas. The unequal fight between poor indigenous communities and oligarchs backed by heavily armed mercenaries and state security forces continues. But, as Berta said, 'We must undertake the struggle in all parts of the world, wherever we may be, because we have no other spare or replacement planet. We have only this one, and we have to take action.'[5]

Worth Reading

Lakhani, N. (2020). *Who killed Berta Cáceres?* New York: Verso.

Worth Watching

Berta Vive (2016)

---

[5] Quoted in Watts, J. (2015): Honduran Indigenous Rights Campaigner Wins Goldman Prize. The Guardian, April the 20th, 2015.

Don't Waste It! – Rossano Ercolini, Bea Johnson, and Ian Kiernan

The most typical invention of our species is waste. Not only are we inclined to waste time and energy, but we also produce huge quantities of household waste, industrial waste, toxic waste, nuclear waste, and all sorts of waste. Incinerators are operating (and polluting the environment), dumping grounds are growing, and garbage patches (i.e., plastic islands) have formed in the oceans, indicating that, as yet, humans did not find a solution to waste management. As businessman and environmental activist Ray Anderson put it, 'Today the throughput of the industrial system, from mine and wellhead to finished product, ends up in a landfill or incinerator. For every truckload of product with lasting value, thirty-two truckloads of waste are produced. That's mindblogging, but it's true. So we have a system that is a waste-making system.'[1] It is no wonder some are fed up with the situation and want to change it.

Working as a teacher in the Italian settlement Capannori, Rossano Ercolini encouraged pupils to recycle paper and plastic, and persuaded the management to use durable instead of disposable utensils in the school canteen. In 1994 local authorities released plans to build an incinerator in the town's vicinity. Ercolini opposed the incinerator due to the expected health risks. He invited experts and arranged meetings and discussions to inform residents about the hazards of waste incineration and the advantages of recycling and composting. Soon there was considerable opposition to the construction of the incinerator. Ercolini organised protests, and after a while the construction plans were cancelled. Ercolini was charged with solving the settlement's waste management.

Ercolini had a vision where landfills and incinerators are not needed any more, a strategy known as zero waste. Ercolini worked with local residents to find solutions that function well and are acceptable for the community. A door-to-door separate collection system was invented, where recyclable material is collected from the resi-

---

[1] Ray Anderson in the 2007 movie 11th Hour.

© Springer Nature Switzerland AG 2019
L. Erdős, *Green Heroes*, https://doi.org/10.1007/978-3-030-31806-2_40

dents' homes. In addition, strong incentives were introduced to encourage people to produce less non-recyclable waste. A great emphasis was placed on informing people why recycling is important. A study has revealed that 99% of the population is participating in the recycling scheme and 94% of them is satisfied with the system. A campaign was started to popularise tap water instead of bottled water, and plastic bottles were banned from all schools and public buildings in Capannori. Negotiations were started with companies to use less packaging material. Inhabitants received composters free of charge and were educated how composting works. In some shops, local products are available in bulk, and consumers can use their own containers – making packaging unnecessary. Within a few years, waste generation was reduced by almost 40%, while the recycling rate increased from 11% to 82%. Thus, the amount of waste ending up in landfills is now only a tiny fraction of what it was a few years earlier. But Capannori was determined to push further, and adopted a zero waste strategy in 2007.

Ercolini's anti-incinerator campaign and the zero waste strategy spread across Italy and the rest of Europe: some incinerators have been shut and plans to build new ones have been abandoned. The idea of zero waste became accepted in several European cities and towns. In 2013 Ercolini was awarded the Goldman Environmental Prize. He is currently the president of the Board of Zero Waste Europe.

The final goal of the zero waste strategy is that no waste has to be dumped or incinerated. Critics say it is a utopian vision and regard the idea as unrealistic. However, it is clear that, in the long run, zero waste is the only acceptable and realistic strategy on a finite planet. But zero waste is more than just a strategy for survival. It is also a philosophy. As Rossano Ercolini says, 'When you throw something away, or, worse, when you are making something to throw away, you are throwing the future away.'[2]

The 3 Rs
According to Ray Anderson, the take-make-waste philosophy of our times ('digging up the earth and converting it to products that quickly become waste in a landfill or an incinerator'[3]) may be regarded as a crime against the future generations. An excellent and easy-to-understand summary of the situation is provided by Annie Leonard in the short documentary *The Story of Stuff* and some related short videos. The problem is now widely recognised and the 'reduce, reuse, recycle' strategy (commonly known as the 3 Rs) has been proposed as a solution.

(continued)

---

[2] Rossano Ercolini in Connett, P. H. (2013): The Zero Waste Solution: Untrashing the Planet One Community at a Time. Chelsea Green Publishing, White River Junction, p. 113.
[3] Ray Anderson's TED talk: https://ted2srt.org/talks/ray_anderson_on_the_business_logic_of_sustainability

To reduce the need for material input (i.e., to take less from nature) is the first and probably the most important part of the strategy. We should buy only those products which we really need. Commercials want us to believe that we need new smartphones every few months, and we are inclined to think some brand new gadgets will make us happy and successful. Stimulated by carefully planned advertisements many believe that buying a new item is better than having the old one repaired. Plastic bags and tin cans are thrown away after a few minutes (or a few seconds) use, representing prime examples of wasting energy and materials. If we asked ourselves whether we really need all these objects, the response, most of the times, would be negative. It seems clear that we should consume less and live more, by re-discovering the value of human relations and meaningful pastimes.

To reuse means that things we no longer need may be used by others, provided that they are in good condition (clothes that do not fit any more, books we will never read again, toys that became needless after the kids grew up, etc.). In addition, objects we do not want to use according to their original function may be used for another purpose.

To recycle includes turning wastes into raw materials, which then can be used to make new products. In several countries recycling is organised and available for everyone. Here, sorting and collecting waste separately is a relatively easy means to reduce our ecological footprint.

Designer William McDonough points out that the quality of the material is reduced during recycling. Thus, recycling, in the form it is performed today, should be termed downcycling. Notwithstanding that downcycling is better than landfills or incinerators, real recycling would mean that products are designed for 100% recycling, without quality reduction. To achieve this goal, design itself should be re-designed.

Bea Johnson was born in France but moved to the US, where her family lived an affluent life in a large house. However, something was missing. When they moved to a smaller house they realised that a simpler life means a life that is richer in experiences and valuable time spent together. It also became clear to them that most stuff they had was unnecessary. They started to understand how harmful their everyday practices had been to the environment. Therefore, led by Bea, the family gradually adopted a new lifestyle and started to sharply reduce their garbage output. As incredible as it may seem at first glance, Bea Johnson managed to reduce the waste of her family to a mere one litre per year. The new life enabled them to spend less money, live healthier, have more free time, and experience more fun together. Bea started a blog about her experiences, published a book that has been translated into over twenty languages, and she regularly gives talks to inspire others to make a transition to a happier, healthier, and greener life.

Ian Kiernan is a builder whose passion has been sailing. He participated in several competitions, including the 1986/87 BOC Challenge, a round-the-world single-handed yacht race. He was shocked by the rubbish he observed everywhere during the voyage. One of his dreams had been to visit the Sargasso Sea, but when he got there he was terribly disappointed because of the pollution. He became so angry that he decided to do something about the sad situation. He thought it would be reasonable to start a campaign in his native Sydney. So, in 1989, with the help of some friends he organised an event called Clean Up Sydney Harbour. Unexpectedly, some 40,000 people showed up and an astonishing 5000 tonnes of rubbish was collected. In the next year the event evolved into the nationwide Clean Up Australia Day, which mobilised about 300,000 people. The project was endorsed by the United Nations and became global when Clean Up the World was launched in 1993. Nowadays more than 30 million people from 130 countries participate each year.

Of course, the Clean Up project is not only about collecting rubbish. The idea includes encouraging people to recycle, compost, save energy, use rechargeable batteries, and refuse plastic bags. Kiernan also urges companies and legislators to make efforts to reduce packaging and improve energy efficiency.

In 1994 Kiernan was named Australian of the Year. In 1998, he received the Sasakawa Environment Prize of the United Nations Environment Programme.

Like Rossano Ercolini, Bea Johnson, and Ian Kiernan, all of us can make a difference. Let's be happy without buying and throwing away compulsively!

Worth Reading

Johnson, B. (2013). *Zero waste home: The ultimate guide to simplifying your life by reducing your waste*. New York: Scribner.
McDonough, W., & Braungart, M. (2002). *Cradle to Cradle: Remaking the way we make things*. New York: North Point Press.

Worth Browsing

www.cleanuptheworld.org
zerowasteeurope.eu
zerowastehome.com
www.mcdonough.com
www.cleanup.org.au

Worth Watching

The Story of Stuff (2007)
The Story of Bottled Water (2010)
The Story of Cosmetics (2010)
The Story of Electronics (2011)
Bag It! (2010)

Al Gore – The Climate Crusader

Politicians who are widely respected among liberal, conservative, and independent citizens throughout the world are few and far between. And only a handful of them are adored like famous musicians or movie stars. Former US Vice President Al Gore certainly belongs to this small group.

Al Gore was born as the second child of Democrat Congressman Albert Gore. The young Al Gore developed a respect for nature at their family farm in Tennessee, and he became aware of environmental problems when he discussed Rachel Carson's book *Silent Spring* with his mother and his sister. Al Gore thinks this book had a profound effect on his worldview. His commitment to environmental issues solidified when he was a student at Harvard University and took a course with Roger Revelle. Revelle was among the first scientists to note that the use of fossil fuels may result in global warming. After graduating in 1969, Gore served in Vietnam as a military journalist. Although Gore opposed the war, he did not want to avoid draft, because he thought someone else would have to go in his place. He served as Congressman from 1977 to 1985, and as Senator from 1985 to 1993. During this period he organised congressional hearings on climate change and the effects of toxic waste, and urged immediate actions to protect the ozone layer. He intensively studied environmental problems and potential solutions, and in 1992 published *Earth in the Balance*.

Earth in the Balance explores multiple aspects of the environmental crisis, including global warming, biodiversity loss, waste management problems, water and soil pollution, and deforestation. The book describes how humans are damaging the planet and destroying their own future. Gore provides readers with shocking and distressing facts, but, in addition to analysing the problems, he also investigates their causes by disentangling intricate networks of human society and the natural world. The author places the current environmental problems in a cultural and psychological context. He criticises the erroneous mechanisms of our societies, and offers solutions to correct them. Al Gore also scrutinises individual and collective responsibility and finds that personal action, political action, and business action has to be taken simultaneously. To address the current global environmental crisis,

© Springer Nature Switzerland AG 2019
L. Erdős, *Green Heroes*, https://doi.org/10.1007/978-3-030-31806-2_41

human mentality has to be changed: the relationship between civilisation and the biosphere has to be rebuilt. Environmental protection should become a central organising principle in decision making. In *Earth in the Balance* Al Gore proposes a global environmental plan to help poor nations reach economic development without depleting natural resources and destroying the environment. The plan follows the idea of the Marshall Plan, the economic aid that helped rebuild much of Europe after World War Two. Today's reader of *Earth in the Balance* may wonder how a politician could be so well-informed about environmental issues. And today's reader certainly feels anger because of the little progress that has been made since the book was published.

From 1993 to 2001 Al Gore was the 45th Vice President of the United States. Bill Clinton selected Al Gore partly because of his environmental commitment. Climate change remained in the focus of Gore. He was instrumental in setting the stage for the Kyoto Protocol, which aimed at a reduction of greenhouse gas emissions. The treaty, however, was eventually not ratified by the USA due to the Senate's resistance.

In 2000 Al Gore ran for president. In one of the most controversial elections in the history of the US, more citizens voted for Gore than for Bush, but Gore narrowly lost in the Electoral College. After the defeat he devoted all his time and energy to environmentalism, developing a slideshow about climate change. The slideshow, which he has presented in many parts of the world, was adapted into a documentary by director Davis Guggenheim in 2006. The film *An Inconvenient Truth* became one of the most successful documentaries of all times: it performed well at the box office and received two Oscars and plenty of other awards. The documentary has a quite unusual structure, linking climate change information with turning points of Al Gore's life. It represents a perfect mixture of popular science and inspirational source, correct but not overloaded with data, convincing, easy to understand, seasoned with humour, and accompanied by great music. The film has been widely applied for education purposes in many countries, and can also prove useful for those who want to improve their presentation skills.

Global Warming in a Nutshell

The Earth's atmosphere is made up of 78% nitrogen, 21% oxygen, and some 0.9% argon. All the other components have very low concentrations. However, some of them play an extremely important role in climate regulation. Carbon dioxide, methane, water vapour, dinitrogen oxide, and chlorofluorocarbons (CFCs) are able to absorb heat energy emitted from Earth's surface and reradiate it to Earth. These gases are called greenhouse gases. Greenhouse effect is a natural phenomenon, as greenhouse gases (with the exception of CFCs) are natural components of the atmosphere. Without the greenhouse effect the Earth would be a rather hostile planet, as its surface temperature would be 33 °C lower than it is today. However, human activity has drastically increased the amount of greenhouse gases in the atmosphere, resulting in global warming. The major anthropogenic sources of carbon dioxide are the burning of fossil fuels and deforestation, while methane is emitted through agricultural processes, mainly livestock breeding.

(continued)

During the last 150 years, the global temperature has increased by ca. 1 °C. If greenhouse gas emission is not reduced considerably, a further increase of up to 5 °C may follow within the next couple of decades. This may not sound too much, but such changes are quite extreme and will have catastrophic consequences.

Extreme weather events will be common: heat waves, prolonged droughts, and heavy rains will be more frequent. Some areas will suffer from floods, others will turn extremely dry, making agriculture impossible. Hurricanes will become stronger. As glaciers disappear, there will be a shortage of drinking water for at least one billion people. Some species will move to higher altitudes or latitudes, others will go extinct, possibly causing ecosystems to collapse. Ocean wildlife is also at risk: it is expected that most coral reefs will die within a few decades. In the Arctic region there will be no summer ice, which will be a profound change for animals that rely on ice, such as polar bears. The melting of glaciers and the ice sheets of Greenland and Antarctica will result in rising sea levels. Low lying areas, including agricultural lands, cities, and several islands will be flooded.

There is strong evidence that the current climate change is a result of human activity. This means we are morally obliged to take action to minimise harm. Governments, corporations, non-governmental organisations, and individual citizens all have to take action. Perhaps the most important thing is to save energy. Consuming energy means producing greenhouse gases. Use electronic devices only if you really need them. Insulate your home. Turn down the thermostat a couple of degrees in winter, and turn it up a bit in summer. Do not waste water. Use public transport, ride a bike, or walk, instead of driving a car. Try to buy products from your own region rather than items that have been produced far away. Prefer environmentally friendly products even if they cost a bit more. The number of things you can do to fight global warming is almost infinite. Reduce your carbon footprint and let the Earth breathe!

In 2007 Al Gore organised *Live Earth*, a 24-hour concert, the aim of which was to raise global environmental consciousness. The events took place on all continents, including Antarctica, where a band formed by scientists of the Rothera Research Station gave a concert. Though *Live Earth* got considerable media attention and thus clearly had the chance to bring the issue to a huge audience, it has been criticised because of its large energy demand and consequent environmental impact. The net effect of such an event is difficult to estimate, but it is sure that questions like this must not be neglected; efforts should be made to minimise and offset environmental impact.

After the Oscars of *An Inconvenient Truth* in 2007, the greatest accomplishment of that year was yet to come. The Intergovernmental Panel on Climate Change

(IPCC) and Al Gore were awarded the Nobel Peace Prize 'for their efforts to build up and disseminate greater knowledge about man-made climate change, and to lay the foundations for the measures that are needed to counteract such change.'[1] Al Gore donated the prize money to the Alliance for Climate Protection, a non-profit environmental organisation he had founded (today it operates under the name The Climate Reality Project).

In his book *Our Choice*, published in 2009, Al Gore focuses on how to solve the challenge of global warming. Though at times most of us may despair over the situation, Al Gore thinks we should not give up hope. 'We can solve the climate crisis. It will be hard, to be sure, but if we can make the choice to solve it, I have no doubt whatsoever that we can and will succeed.'[2]

An Inconvenient Sequel was released in 2017. Not only is it inconvenient, but it is a shocking movie and surely one of the best follow-ups ever made.

In addition to the environment, Al Gore shows vivid interest in many topics at the intersection of society, communication, science, and technology. At Harvard he wrote his thesis about the effects of television on presidency. In his 2007 book *The Assault on Reason* he investigates a similar topic. In modern America (and in several other countries as well) real discussions have almost totally disappeared from politics. Reason has been replaced by propaganda, discussion by demagoguery, truth by manipulation, fact by deception, and information by disinformation. Gore explores why this has happened, how it distorts democracy, which fatal consequences it may have, and how it can be counteracted.

In another book entitled *The Future*, Al Gore analyses the present situation and the prospects of mankind and our planet. Achievements of science, technology, and economics are incredible, but the dark sides cannot be neglected either. We are at a crossroads. Either we use accumulated knowledge in order to move toward equity, fairness, and sustainability, or we opt for another direction and destroy our own future. *The Future* is a valuable resource, but deserves serious criticism due to its views regarding experiments using animals and human embryos, and because of the book's overly US-centred perspectives.

In a 2007 interview Gore said that environmental efforts had thus far failed. He added the followings: 'I think we're making progress; it's just that nothing has matched the scale of the response that is truly needed.'[3] Despite being a Nobel laureate and one of the best known and most respected environmental activists, Al Gore is dissatisfied: 'Until we start sharply reducing global-warming pollution, I will feel that I have failed.'[4] May we hope that humanity will eventually find the way to a clean future.

---

[1] www.nobelprize.org/prizes/peace/2007/summary/

[2] Gore, A. (2009): Our Choice: A Plan To Solve the Climate Crisis. Rodale, New York, p. 15.

[3] Interview with Al Gore, Time, December the 19th, 2007. content.time.com/time/specials/2007/personoftheyear/article/0,28804,1690753_1695388_1695516,00.html

[4] Interview with Al Gore, Time, December the 19th, 2007. content.time.com/time/specials/2007/personoftheyear/article/0,28804,1690753_1695388_1695516,00.html

Worth Reading

Gore, A. (2006a). *An inconvenient truth: The planetary emergency of global warming and what we can do about it*. New York: Rodale.

Gore, A. (2017). *An inconvenient sequel: Truth to power*. New York: Rodale.

Gore, A. (2006b). *Earth in the balance: Ecology and the human spirit*. New York: Rodale.

Gore, A. (2009). *Our choice: A plan to solve the climate crisis*. New York: Rodale.

Gore, A. (2007). *The assault on reason*. New York: The Penguin Press.

Gore, A. (2013). *The future: Six drivers of global change*. New York: Random House.

Worth Browsing

www.algore.com
www.climaterealityproject.org
www.co2.earth

Worth Watching

An Inconvenient Truth (2006)
An Inconvenient Sequel (2017)
Glacial Balance (2013)

Arnold Schwarzenegger – The Strong Man of Environmentalism

Anything is possible in a Hollywood movie. But reality surpasses imagination and produces stories that could never happen in a screenplay. Once in history an Austrian boy can become a bodybuilding champion, an action hero, governor of California, and a leading environmentalist. A ruthless terminator may turn into green Governator.

Arnold Schwarzenegger was born in 1947 in Thal, a small village in the Austrian Alps. He started bodybuilding as a teenager and won Mr. Universe title at the age of 20. He moved to the USA in 1968 and continued his bodybuilding career, winning several competitions, including seven Mr. Olympia contests. Schwarzenegger moved into acting, and gained worldwide fame with the 1982 movie *Conan the Barbarian*. He became an icon with his roles in films such as *The Terminator* and *Predator*, now widely considered classics of motion picture history.

Schwarzenegger's philanthropic work has been diverse from the 1980s. He participated in an anti-drug campaign, has supported the Special Olympics and children's after-school programs, and was a Red Cross ambassador. He has increasingly been involved in politics from the 1990s: he served as Chairman of the President's Council on Physical Fitness and Sports and as Chair of the California Governor's Council on Physical Fitness and Sports.

Schwarzenegger traces back his environmental advocacy to his childhood, when his family lived a simple and frugal life set in Austrian countryside. But it was his 2003 gubernatorial campaign when his attention actually turned to the environment. Schwarzenegger was elected 38th Governor of California and served two terms. Initially most environmentalists were rather sceptical and thought Schwarzenegger's green words had been empty campaign promises, but their doubts proved impetuous. As journalist Ronald Bronstein wrote: 'on questions surrounding the transition toward a new energy economy, no governor would prove as visionary or determined.'[1] Schwarzenegger took significant steps to reduce California's greenhouse gas emissions. The 2006 Global Warming Solutions Act is among his major accomplishments.

---

[1] Brownstein, R. (2009): The California Experiment. The Atlantic, October 2009. https://www.theatlantic.com/magazine/archive/2009/10/the-california-experiment/307666/

© Springer Nature Switzerland AG 2019
L. Erdős, *Green Heroes*, https://doi.org/10.1007/978-3-030-31806-2_42

Schwarzenegger also fought for stricter fuel-efficiency standards, and, though the car manufacturer lobby tried to resist, Californian ideas were eventually accepted as models for federal regulations. Schwarzenegger initiated a network of hydrogen filling stations and a large-scale installation of solar panels across the state. His leadership on sustainability contributed to an increase in Californian energy efficiency and renewable use. As a governor, Schwarzenegger also took action for conserving endangered Californian species and ecosystems, both terrestrial and marine. Though much remains to be done, it is clear that California is moving toward a low-carbon future.

Schwarzenegger is an animal lover and a fierce opponent of trophy hunting. 'Take a photo, not a shot'[2] he says. In an action-styled video made for the Wildlife Conservation Society's anti-poaching campaign Arnie calls for an end of killing elephants. He also backed National Geographic's Big Cats initiative. After Cecil, the most famous and beloved lion of the Hwange National Park in Zimbabwe was killed by an American dentist, Schwarzenegger asserted that 'Killing a lion isn't ballsy.'[3] On Instagram he posted a photo of himself with some prizes in his hands, intended as a visual definition of what is a real trophy, adding that lions should never be regarded as trophies.

Arnold Schwarzenegger appeared with Bill Nye in *Global Meltdown*, one of National Geographic's most unconventional documentaries, which explores climate change from a fresh perspective. Arnie narrated the 3D nature film *Wonders of the Sea*, directed by Jean-Michel Cousteau and Jean-Jacques Mantello.

It is well-known that livestock breeding contributes significantly to global warming, thus cutting down global meat consumption is a necessary part of the fight against global climate change. This is why Schwarzenegger advocates reducing meat consumption and co-operated with director James Cameron to persuade people to eat less meat. As their slogan goes 'less meat, less heat, more life.'[4]

Schwarzenegger characterises himself as a moderate Republican, and has heavily criticised Presidents Bush and Trump because of their extremely poor performances in environmental protection. Schwarzenegger believes in a new way of politics, which is based on co-operation. Division, a normal part of old-way politics, makes no sense in environmentalism. As Schwarzenegger keeps saying, 'There is

---

[2] Arnold Schwarzenegger on Twitter. Quoted in Jancelewicz, C. (2016): Arnold Schwarzenegger's elephant close encounter: 'Take a photo, not a shot.' Global News, May the 31st, 2016. globalnews. ca/news/2731718/arnold-schwarzeneggers-elephant-close-encounter-take-a-photo-not-a-shot/

[3] Arnold Schwarzenegger on Twitter. Quoted in Brown, K. (2015): 'These are trophies, Cecil the lion is not,' says Arnold Schwarzenegger. The Telegraph, July the 31st, 2015. www.telegraph.co. uk/film/movie-news/arnold-schwarzenegger-cecil-the-lion/

[4] www.youtube.com/watch?v=GTyo7qOHJJo, see also: Shoard, C. (2016): Arnold Schwarzenegger and James Cameron urge people to eat less meat. The Guardian, June the 23rd, 2016. www.the-guardian.com/film/2016/jun/23/arnold-schwarzenegger-james-cameron-eat-less-meat-china

no liberal air or conservative air. We all breathe the same air.'[5] Clean air, safe drinking water, healthy soils, and liveable settlements should be equally important for everyone, independent of their political thoughts.

Schwarzenegger also has a few words for those who doubt the reality of human-caused global warming. There is a very strong consensus among climate scientists that global warming is happening and that it is caused by humans. However, even if all these

Climate Change Denial

Climate change denial (also called global warming scepticism) is the view according to which measures taken to reduce greenhouse gas emissions are unnecessary or even harmful to society. The exact position taken by sceptics is usually hard to decipher. Some of them state that global climate is not changing currently. Others accept that climate shows a warming trend, but deny that it is caused by human activity. A third group of sceptics accepts the reality of anthropogenic (i.e., human-caused) warming, but thinks it will have no negative effects. The overwhelming majority of sceptics does not have any training in environmental sciences, simply meaning that they do not understand what they are talking about. Some sceptics have financial interests in the fossil fuel industry, thus they oppose environmental measures because these would reduce their profits.

Climate change denial rests on a number of misunderstandings, misinterpretations, myths, and outrageous lies. Temperature records from thousands of measuring stations all around the world unequivocally show that the global temperature is rising. Measurements carried out by satellites produce the same result. Humans increase the amount of greenhouse gases in the atmosphere; not even sceptics deny that. Physical mechanisms of how greenhouse gases cause increasing temperature are well understood. Alternative explanations (e.g., variations in solar activity or cosmic rays) can be ruled out as possible sources of the warming. Climate models also indicate that human activity is responsible for the current climate change. The consequences of continuing warming may be catastrophic. Climate change is likely to trigger an extinction wave and eco-system collapse, as evolutionary processes and species' migrations cannot keep pace with rapidly increasing temperatures. Though it is predicted that global warming may benefit human societies in certain regions in the short-term, these will be outweighed by the negative impacts of climate change.

Scepticism is an essential part of science. But climate change denial is not real scepticism, neither is it science. Rather, climate change denial belongs to the realm of pseudo-science and is akin to alchemy, horoscope, and the flat Earth theory.

---

[5] Quoted in Murray, J. (2017): Macron Meets Schwarzenegger and Vows to Stop Oil and Gas Licences. The Guardian, June the 26th, 2017. www.theguardian.com/environment/2017/jun/26/macron-meets-schwarzenegger-vows-stop-oil-gas-licences

scientists were wrong, we still would have a moral duty to reduce and then stop fossil fuel combustion. Burning fossil fuels does a lot of harm independent of climate change, including acid rain and serious respiratory diseases. Millions of people die each year because of fossil fuel. It seems clear that we should switch to alternative sources.

Scientists agree that combatting climate change requires immediate action. We have an obligation to urge necessary measures at the national and international levels, but if they won't happen, we should proceed at sub-national levels such as federal states, provinces, counties, and cities. This is the idea behind R20 – Regions of Climate Action, a non-profit organisation founded by Schwarzenegger. R20's goal is to implement projects that increase sustainability by connecting authorities with reasonable technologies and investors.

Arnold Schwarzenegger's life is a proof that nothing is impossible, provided that one works hard enough to achieve their goals. When coupled with passion and determination, optimism does make sense, and this also applies to environmental issues. The former governor does not deny that the problems we face are tremendous, but he strongly believes they are solvable. Environmentalism should be perceived as a challenge rather than a nuisance. If implemented properly, environmental measures are good for the economy as well, as they stimulate innovation and create green jobs.

Schwarzenegger thinks the same positive attitude should be used when communicating sustainability. Scientific facts, hard data, dark predictions and warnings certainly have their role in informing people, but placing greater emphasis on positive messages may prove fruitful. Environmentalism can become sexy if we emphasise that the use of existing green technologies is easy, reducing our carbon footprint is trendy, living a simpler life makes a lot of fun, and, at a more general level, environmentalism brings happiness to one's life.

We have a lot to do, and time is running out. But there is still hope that we can terminate the destruction of our environment!

Worth Reading

Schwarzenegger, A. (2013). *Total recall: My unbelievably true life story*. New York: Simon and Schuster.

Worth Browsing

regions20.org
schwarzenegger.com

Worth Watching

Bill Nye's Global Meltdown (2015)
Wonders of the Sea (2017)

Greening Hollywood – The Activism
of Leonardo DiCaprio

Celebrity involvement in various good causes is flourishing in Hollywood. But few of the stars are as dedicated as Leonardo DiCaprio is to nature conservation, animal advocacy, and environmental protection.

Leonardo DiCaprio was born in 1974. In his early career he appeared in television commercials and series. In the 1990s he starred in movies like *This Boy's Life* and *Romeo+Juliet*. The breakthrough came with James Cameron's 1997 blockbuster *Titanic*, which catapulted DiCaprio to global stardom. Among his most memorable roles were those in *The Great Gatsby*, *The Wolf of Wall Street*, and *The Revenant*. For this latter, he received an Academy Award.

As a kid Leonardo DiCaprio had a vivid interest in the natural world. But his actual activism was instigated and inspired by his 1998 conversation with Vice President Al Gore. In that year DiCaprio established the Leonardo DiCaprio Foundation with the mission of 'ensuring the long-term health and well-being of all Earth's inhabitants.'[1] The Foundation has two basic approaches. The first is to raise money and donate it to environmental and conservation organisations. As of 2018 the Foundation has supported over 200 projects with a total of $100 million. Their second auction alone, hosted by Leonardo DiCaprio, raised over $40 million. One of the key areas where the Leonardo DiCaprio Foundation has been most active is the protection of indigenous peoples' rights, supporting the fight of native tribes to protect their land and culture from being devastated by oil or gas extraction and palm oil plantations. The Foundation collaborates with WWF to save tigers and with WildAid to protect sharks, and is involved in numerous other projects connected to ocean and forest protection and climate change mitigation. The second approach is to initiate public campaigns. For example, a petition backed by the Leonardo DiCaprio Foundation was signed by some 1.5 million people and persuaded the government of Thailand to take steps to end ivory trade. The Chief Executive Officer of the Leonardo DiCaprio Foundation is Terry Tamminen, who worked with Governor Schwarzenegger to make Californian energy usage more sustainable.

---

[1] www.leonardodicaprio.org/about/

© Springer Nature Switzerland AG 2019
L. Erdős, *Green Heroes*, https://doi.org/10.1007/978-3-030-31806-2_43

DiCaprio was chairman of the Earth Day 2000 celebrations, during which he participated in an ABC special and interviewed President Clinton about the status and future prospects of the environment.

The 2000 movie *The Beach*, starring DiCaprio, received heavy criticism from environmental groups, because it had ruined the natural habitats at the filming location in Thailand. When the shooting was over, DiCaprio contacted conservation organisations to support the ecological restoration of the area. DiCaprio has worked intensively ever since to make movies more environmentally friendly.

Leonardo DiCaprio has done a great job in joining his love of the environment and expertise in film industry. He created and narrated *Global Warning*, a short documentary about the reality of climate change. The programme was inspired by Thom Hartmann's book *The Last Hours of Ancient Sunlight*. A similar short film entitled *Water Planet* was written and narrated by DiCaprio. Leo was the narrator of the short documentaries *Carbon*, *Last Hours*, *Green World Rising*, and *Restoration*.

In 2007 DiCaprio produced, co-wrote and narrated *The 11th Hour*, an excellent environmental documentary that featured interviews with leading scientists and environmentalists, among many others Stephen Hawking, Wangari Maathai, David Suzuki, Lester Brown, Stephen Schneider, Herman Daly, David Orr, Paolo Soleri, Andrew Revkin, and Wes Jackson. The movie explored the question whether individual environmental problems are separate or are parts of a deeper global crisis. *The 11th Hour* received positive reviews from critics and is usually considered one of the most influential films about environmental matters.

DiCaprio served as executive producer for a string of films connected to green issues. The 2014 movie *Virunga*, which was nominated for an Academy Award for Best Documentary Feature, is a shocking story about real heroes who risk their lives to protect mountain gorillas amidst war and unscrupulous economic exploitation. No less shocking is *Cowspiracy*, a movie that unveils the destructive effects factory farming has on natural habitats and human environment. *Catching the Sun* scrutinises how the transition to clean energy affects the economy and the society. *The Ivory Game*, an action-styled, thrilling documentary, uncovers the illegal network behind elephant poaching and warns that elephants could become extinct within the next 15 years if the demand continues to support the multi-million dollar ivory business.

In 2016 Leonardo joined forces with National Geographic and produced *Before the Flood*, an extremely successful documentary about climate change. The movie follows DiCaprio's journey to witness the consequences of global warming as he discusses the topic with scientists, activists, and world leaders from Barack Obama to Pope Francis. 'We wanted to create a film that gave people a sense of urgency, that made them understand what particular things are going to solve this problem,'[2] DiCaprio said about the primary aim of the film.

---

[2] Quoted in Hickman, L. (2016): 7 Key Scenes in Leonardo DiCaprio's Climate Film 'Before the Flood.' EcoWatch, October the 21st, 2016. www.ecowatch.com/leonardo-dicaprio-before-the-flood-2057070140.html

Pollinators such as bees and butterflies are key components of ecosystems. Their work is indispensable as most of our agriculture would be virtually impossible without them. Yet pollinators are declining worldwide due to the spread of monocultures and the use of pesticides. The short documentary *Pollinators under Pressure*, narrated and executive produced by Leonardo DiCaprio, informs people about what they can do to relieve the situation.

The most recent environmental film related to DiCaprio is *Ice on Fire*. The documentary, produced and narrated by DiCaprio, investigates how we can reverse global warming.

Leonardo DiCaprio has served on the boards of several organisations such as the World Wildlife Fund, the Natural Resources Defense Council, and the International Fund for Animal Welfare, and he was appointed as United Nations Messenger of Peace on Climate Change. For DiCaprio, participating in protests is also a necessary part of activism, so he has repeatedly joined demonstrations demanding immediate measures against climate change.

DiCaprio uses every opportunity to raise environmental awareness. His speech before the 2014 Climate Leaders Summit made news worldwide, and reached an estimated one billion people around the globe. When receiving his third Golden Globe, he spoke about the rights of indigenous people. He even used his Academy Award acceptance speech to draw attention to global warming: 'Climate change is real, it is happening right now. It is the most urgent threat facing our entire species, and we need to work collectively together and stop procrastinating. We need to support leaders around the world who do not speak for the big polluters, the big corporations, but who speak for all of humanity, for the indigenous people of the world, for the billions and billions of underprivileged people who would be most affected by this, for our children's children, and for those people out there whose voices have been drowned out by the politics of greed.'[3]

Supporting the campaign of the Animal Legal Defense Fund, Leo spoke out for Tony, a tiger who was kept at a truck stop in Louisiana. Tony lived alone in a very small cage, inhaling exhaust fumes for his entire life. The campaign demanded that Tony be relocated to an accredited sanctuary. Unfortunately, the animal lovers' efforts failed and Tony died in 2017. The owner of the truck stop is planning to buy a new tiger to replace Tony, but the Animal Legal Defense Fund will do everything to prevent him from doing so.

Leonardo DiCaprio has made efforts to reduce his own environmental impact. He is a vegetarian, uses solar energy, usually rides a bike, and was among the first stars to buy a hybrid car. He offsets his carbon emissions by donating to an environmental organisation that plants trees. Moreover, DiCaprio uses commercial flights instead of private jets whenever possible. This may not seem a large sacrifice for most of us, but for an A-list Hollywood movie star it is quite a significant step.

---

[3] Quoted in Hoffman, A. (2016): How Leonardo DiCaprio Got People to Care About Climate Change. Time, August the 5th, 2016. time.com/4441219/leonardo-dicaprio-oscars-climate-change/

As a young boy Leonardo DiCaprio wanted to be either an actor or a marine biologist. He has conquered the world as a movie star, but his contributions to environmental protection, nature conservation, and animal advocacy match up to his achievements in acting. Over the last two decades DiCaprio has repeatedly demonstrated that he is one of the most passionate and most committed green activists in Hollywood. He believes in the causes he is fighting for and really wants to make a difference. And he encourages all of us to take action. 'We are running out of time and it is now incumbent upon all of us, all of you, activists, young and old, to please get involved. Because the environment and the fight for the world's poor are inherently linked. The planet can no longer wait, the underprivileged can no longer be ignored. This is truly our moment for action.'[4]

The Greta Effect

Greta Thunberg represents the youngest generation of green activists. Born in 2003, the Swedish environmental campaigner was shocked when she learned about climate change. She tried to reduce her environmental impact and persuaded her family to do the same. But she wanted to do more. She thought that going to school and learning scientific facts does not make sense if politicians are ignoring scientists' warnings about climate change. So, in 2018 she started a school strike for climate: instead of going to school, she sat down in front of the Swedish parliament, demanding immediate action to fight climate change. At first she was alone but later more and more people joined her. After a couple of weeks Greta continued to strike only on Fridays. The campaign attracted huge media attention and developed into the movement 'Fridays for Future,' with millions of participants, young and old, all over the world. Greta's immense impact has been described as 'the Greta effect.' Her appearance boosted public interest in action against global warming.

Greta Thunberg was nominated for the Nobel Peace Prize, received the prestigious Right Livelihood Award, and was selected Person of the Year 2019 by *Time* magazine. 'I thought, instead of worrying about how future might turn out, you should try to change it while you still can,' Greta said. It seems the world is listening. The only question is if leading politicians are listening.

Worth Browsing

www.leonardodicaprio.org
www.leonardodicaprio.com
www.beforetheflood.com
virungamovie.com

---

[4]Leonardo DiCaprio at the 2015 Global Citizen Festival. www.youtube.com/watch?v=-NN43FV3z5A

theivorygame.com
www.pollinatorsunderpressure.org
www.cowspiracy.com
www.catchingthesun.tv

Worth Watching

The 11th Hour (2007)
Before the Flood (2016)
Ice on Fire (2019)
Virunga (2014)
The Ivory Game (2016)
Cowspiracy: The Sustainability Secret (2014)
Catching the Sun (2015)
Pollinators under Pressure (2018)
Global Warning (2003)
Water Planet (2005)
Carbon (2014)

Epilogue

The green movement has made tremendous progress during the last few decades. As I tried to show in this book, many influential thinkers and activists have joined the cause. However, it would be a mistake to think that their work will be enough to solve the current crisis. Never have we faced greater challenges and the team of green heroes is still too weak. All of us are needed on board.

The condition of our environment is deteriorating rapidly, natural areas and biological diversity are diminishing, tens of billions of animals are tortured and killed each year, and approximately one billion humans are undernourished or have no access to clean drinking water.

You may ask what you can do about all this. First of all, do not turn your head away. Stay informed about the problems and their solutions. Be aware of the challenge but do not despair. Abandon the old and obsolete anthropocentric worldviews. Next, check your lifestyle and try to find ways how you can minimise your negative and maximise your positive impacts. Keep in mind that even the smallest lifestyle adjustments are important. Every little piece of nature conserved and every single life saved counts. Be proud of your results but always try to improve further. At first some actions may seem a bit demanding but the happiness caused by doing what is right will be greater. Moreover, you will realise that ethical behaviour earns respect and even attracts followers. Those who work in the green movement not only make the world a better place, but they themselves become better persons.

Mahatma Gandhi said, 'If we could change ourselves, the tendencies in the world would also change.'[1] Let's start it right now!

---

[1] The Collected Works of Mahatma Gandhi, vol. XII. The Publications Division, Ministry of Information and Broadcasting, Government of India, 1964, p. 158.

© Springer Nature Switzerland AG 2019
L. Erdős, *Green Heroes*, https://doi.org/10.1007/978-3-030-31806-2